O9-AHT-567

CALGARY PUBLIC LIBRARY

523. 47 HUN
Hunt, Garry E.
Atlas of Uranus.
95592222

ATLAS OF URANUS

Atlas of
URANUS

Garry Hunt and Patrick Moore

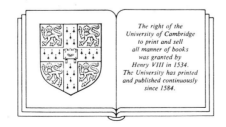

The right of the
University of Cambridge
to print and sell
all manner of books
was granted by
Henry VIII in 1534.
The University has printed
and published continuously
since 1584.

CAMBRIDGE UNIVERSITY PRESS

Cambridge

New York   New Rochelle   Melbourne   Sydney

Published by the Press Syndicate of the University of Cambridge
The Pitt Building, Trumpington Street, Cambridge CB2 1RP
32 East 57th Street, New York, NY 10022, USA
10 Stamford Road, Oakleigh, Melbourne 3166, Australia
© Cambridge University Press 1988

First published 1989

Printed in Great Britain by W.S. Cowell Ltd., Ipswich

British Library Cataloguing in Publication Data

Hunt, G.E.
  Atlas of Uranus.
  1. Uranus (Planet)
  I. Title   II. Moore, Patrick
  523.4'7   QB681

Library of Congress Cataloguing in Publication Data available

ISBN 0 521 34323 2 hard covers

DS

Contents

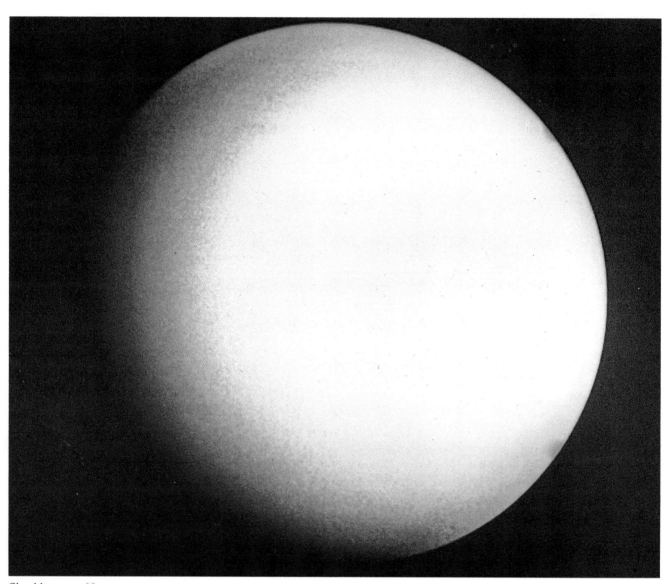

Cloud layers on Uranus

Preface

The aim of this book is to give a picture of the planet Uranus and its system which is as up to date as possible. Our knowledge has been increased immeasurably by the results from the Voyager 2 space-craft, which flew past Uranus in January 1986; we have little hope of learning much more before a new probe makes a rendezvous with the planet, and this will probably not be for several decades at least.

The book was planned by both of us, working closely together, as on previous occasions; but Hunt (one of NASA's Principal Scientific Investigators for the planetary missions) has been responsible for the more technical sections dealing with the space-craft itself, the magnetosphere, the ring-system as revealed by Voyager, and the structure of Uranus, while Moore has tackled the historical and descriptive sections. We can only hope that the result will be of interest.

G.H.
P.M.

Picture acknowledgements

The publishers gratefully acknowledge the help of the following individuals and organisations for allowing us to use their material in this book. Every effort has been made to obtain permission to use copyright materials; the publishers apologise for any errors and omissions and would welcome these being brought to their attention.

6 NASA/JPL; 19, 21, 22, 23 Royal Astronomical Society; 29 Danielson/Tomasko; 33 NASA/Ames; 34 NASA/ JPL; 35 David Allen/AAO; 37 NASA/Science Photo Library; 40, 41, 43, 44, 45, 46, 47, 48, 49, 50, NASA/JPL; 53, CSIRO; 54, 55 NASA; 60, 61, 62, 63, 64, 65, 70, 71, 72, 73, 74, 75, 77, 80, 81, 82, 83, 84, 85, 86, 87 NASA/ JPL.

Illustrations drawn by Paul Doherty and Dataset Marlborough Design Limited, Oxford.

Introduction

Far beyond Saturn, outermost of the planets known in ancient times, there moves the strange, tilted world Uranus. Before the year 1781 it was not only unknown, but quite unsuspected; it had been generally assumed that the Solar System must be complete, and that the vast spaces between Saturn and the nearest star were populated only by wraithlike objects such as comets.

Uranus at its brightest is just visible without a telescope, but only when you know where to look for it – and no Earth-based instrument will show anything definite upon its tiny, greenish disk. So how could we find out what kind of a world it really is? We could analyse its light by means of the spectroscope; we could of course work out its movements with great precision; we could study its family of moons, or at least those which were within telescopic range; and we could make intelligent speculations about its internal make-up. In 1977 a system of dark rings was discovered, quite unexpectedly, and astronomers began to realize that they were dealing with a planet even more peculiar than they had supposed. Yet all in all, our positive knowledge about Uranus was remarkably slight. We were not even certain of the length of its 'day' – the time taken for it to spin upon its sharply-tilted axis.

But also in 1977, on 20 August, a space-probe blasted away from Cape Canaveral, in Florida, to begin a journey which – so far as we can tell – has no ending. It was called Voyager 2, and it was given a programme which many people regarded as over-ambitious. It was scheduled to pass by Jupiter in July 1979 and by Saturn in 1981, sending back close-range information not only about the two giants themselves but also about their satellites. Of special importance was Titan, the senior attendant of Saturn, which has a dense, cloudy atmosphere. Voyager 2 accomplished these objectives almost faultlessly (apart from a temporary breakdown in communication after the Saturn pass). Therefore it became possible to undertake an extra part of the mission – rendezvous with the two outer giant planets, Uranus in January 1986 and Neptune in August 1989. The Neptune pass would be the final encounter. Voyager 2 would never return; it would simply move away into interstellar space, and eventually all trace of it would be lost.

It is doubtful whether many of the space-craft planners dared to hope for complete success, particularly when the Uranus and Neptune encounters were added to the original programme. It would mean a working life of more than a dozen years, and there was so much that could go wrong; an apparently trifling fault could cripple the whole transmitting system, for instance, and bring the mission to an end so far as we were concerned. Moreover, it would be wrong to pretend that the technology of 1977 was nearly as sophisticated as that of a decade later. Progress has been extremely rapid, so that by the time Voyager 2 reached Uranus it was already an 'old' probe.

Yet on the other hand, the whole concept would have seemed absolutely inconceivable only a few years earlier. Before the 1939 war, those who dared to predict journeys to the Moon and flights to the planets were regarded with amiable contempt, and in the eyes of many reputable scientists the British Interplanetary Society and bodies of the same type were regarded as strictly comparable with the Flat Earth Society. It came as a shock when, on 4 October 1957, the Russians launched Sputnik 1, their football-sized satellite which ushered in the Space Age not with a whimper, but with a very pronounced bang.

There followed the long period often called that of the 'race to the Moon', though whether the Russians and the Americans were seriously trying to beat each other to the lunar surface is highly questionable. Meanwhile, the inner planets started to come within range. Venus was by-passed in 1962, when Mariner 2 sent back the rather unwelcome news that instead of being warm and friendly, the Planet of Beauty had a scorching hot surface, a dense, unbreathable atmosphere, and clouds rich in corrosive sulphuric acid, so that it approximates very closely to the conventional picture of hell. Mars was next on the list; Mariner 4 surveyed it in 1964, and by 1976 two American probes, the Vikings, had made controlled landings there and sent back information direct from the bleak Martian landscape. Another Mariner, No.10, made close-range contact with the inhospitable little world Mercury, with its mountains, valleys and craters. Our knowledge of the inner Solar System had grown beyond all recognition in less than two decades.

The outer planets presented much more serious problems, mainly because of their distance. It was a lucky chance that in the late 1970s the four giants – Jupiter, Saturn, Uranus and Neptune – were 'strung out' in a way which meant that it would be feasible to send a space-craft from one to the next in a sort of cosmic tour. Such a chance will not occur again in our time, and the NASA scientists decided to make the most of it.

Both the present authors were at Mission Control, in California, when Voyager 2 made its pass of Uranus at 17 hours 58 minutes 51 seconds GMT on 24 January 1986. Surprises were expected – and they came. What we hope to do, in this book, is to tell the story of Uranus, from the time of its discovery by a German-born amateur working in his garden, using a home-made telescope, to the moment when Voyager 2 achieved the third triumph of its remarkable career.

Uranus in the Solar System

When we look back at the past history of the Solar System, we have at least a few important clues to guide us. First, there is the age of the Earth, which we know with reasonable certainty to be around 4.6 thousand million years. Of course, it is impossible to be precise (unlike the seventeenth-century Archbishop Ussher of Armagh, who pinpointed the creation to within ten minutes on the morning of 26 October, BC 4004!), but there are various lines of research, all of which lead to much the same answer. There is confirmation too, from analyses of the material from the Moon and from meteorites.

Secondly, it also seems highly probable that the planets were formed from a 'solar nebula' – that is to say, a disk-shaped cloud of material associated with the youthful Sun. Moreover, there is a certain orderliness about the Solar System, and its overall arrangement is highly significant.

The System is divided into two very definite parts. First we have the four inner planets – Mercury, Venus, the Earth and Mars – all of which are relatively small (the Earth is actually the largest of them) and have solid surfaces. Beyond the orbit of Mars there is a wide gap, filled by thousands of dwarf worlds known as the minor planets or asteroids; only one, Ceres, is as much as 1000 kilometres in diameter, and the smaller members of the swarm are mere lumps of material, no doubt irregular in shape. Outside the asteroid belt we come to the giants Jupiter, Saturn, Uranus and Neptune, which are very different from the Earth. Their visible surfaces are gaseous, and inside they seem to be mainy liquid, with relatively small solid cores.

The most plentiful element in the entire universe is hydrogen, which is also the lightest of the elements; its atom consists solely of a positively-charged particle termed a proton, round which moves a negatively-charged electron. Hydrogen atoms outnumber the atoms of all other elements put together. The Sun contains a high percentage of hydrogen, and so do the giant planets, but there is much less in the make-up of the Earth or the other inner worlds, and it seems that conditions early on in the story of the Solar System must have been responsible. The planets grew up from the solar nebula by the process of accretion; close to the now-luminous Sun the temperatures were high, so that the lightest elements, notable hydrogen, were driven away. Further out, in cooler regions, the hydrogen could be retained, and the planets in these remote areas had ample opportunity to grow.

Presumably the solar nebula was rotating, and it is very significant that the planets orbit the Sun in the same direction. There is no such thing as a retrograde or 'wrong-way' planet, so far as we can tell, though there are retrograde comets – including Halley's, which caused such interest during its latest return in 1986 even though it never became as spectacular as it has often done in the past. Moreover, the planetary orbits are not very different from circles. The eccentricity of the Earth's orbit is only 0.017. At its closest to the Sun (perihelion, in December) the distance which separates us from the Sun is 147 000 000 kilometres; at its furthest (aphelion, in early July) the distance is increased to 152 000 000 kilometres – but a difference of five million kilometres is not very much. Draw the Earth's orbit to scale, making it fit on a page of this book, and you will barely be able to distinguish it from a circle. Not so with comets, whose paths in most cases are very elliptical. On average, Halley's Comet has a revolution period of 76 years; at perihelion it is closer in than Venus, while at aphelion it wanders well beyond the orbit of Neptune. But comets are flimsy, wraithlike objects, of very low mass. Their only relatively substantial parts – their nuclei – can never be more than a few kilometres in diameter.

Uranus, with an orbital eccentricity of 0.047, has a distance-range from the Sun of between 3004 and 2735 million kilometres. Its revolution period is 84 years; it was last at perihelion in May 1966, when it was at its closest to the Earth on 9 March. Its distance from us was then 17.29 astronomical units – one astronomical unit being the mean distance between the Earth and the Sun (in round figures, 150 000 000 kilometres). It was at aphelion in 1925, and will be so again on 27 February 2009.

Obviously it is a very slow mover; it crawls against the starry background, though its shift is easily detectable from one night to the next when sufficient optical power is used. When a planet is exactly opposite the Sun in the sky, it is said to be at opposition. With Uranus this happens every 370 days.

In 1989 Uranus enters Sagittarius, and remains there until 1995, when it will pass into Capricornus (the Sea-goat). Throughout the whole of the rest of our century, therefore, it will be well south of the celestial equator, so that observing conditions so far as British observers are concerned are poor. In June 1989 Uranus will reach its southernmost point, almost 24 degrees from the equator.

So far we have said nothing about the ninth planet, Pluto, which was not known until its discovery by the American astronomer Clyde Tombaugh in 1930[†] (Neptune had been added to the Solar System much earlier, in 1846). Pluto is an oddity. We now know that it is not only the smallest of the planets, but is even smaller than our Moon, with a diameter of below 2000 kilometres. It seems to be made up of a combination of

---

[†] For the full story see *Out of the Darkness: the Planet Pluto*, by Clyde Tombaugh and Patrick Moore (Lutterworth, UK, and Stackpole, USA, 1981).

rock and ice, so that its mass is very low indeed. Moreover, it has a most unusual orbit – unusual for a planet, that is to say. The revolution period is 248 years; the orbital eccentricity is so high that while for most of the time Pluto moves far beyond any of the other planets, perihelion brings it inside the orbit of Neptune. The next perihelion is due in 1989, and from now until 1999 Neptune, not Pluto, qualifies as 'the outermost planet'.

But is Pluto a genuine planet at all? Evidence today is rather against it. It has a satellite, Charon, one-third the diameter of Pluto itself, and the modern tendency is to regard the two either as a pure-maverick pair or else as a double asteroid. We do know of a smaller body, Chiron (not to be confused with Charon) which spends most of its time between the orbits of Saturn and Uranus, and no doubt there are others.

If Pluto's orbit crosses that of Neptune, we are entitled to ask whether there is any danger of collision. The answer is 'no', at least in the present epoch, because Pluto's orbit is tilted at the exceptionally sharp angle of 17 degrees to that of the main plane of the Solar System, whereas the inclination of the orbit of Neptune is less than 2 degrees. Indeed, it is true to say that in the foreseeable future Pluto is capable of making closer approaches to Uranus than it can do to Neptune.

Let us now consider some of the features of Uranus itself, seen against the general background of the Solar System. It is not always easy to appreciate how spread-out the System is, and a chart tends to give the impression that the four giants are fairly close to each other – just as an unwary European looking at a map of Oceania may fondly imagine that is almost possible to stand on the coast of Queensland and leap across to New Zealand! In fact the distances are vast. Suppose, for instance, that it were possible to climb into some ultra-modern space-craft and fly straight from the Earth to the orbit of Uranus, travelling at a steady rate of a million kilometres per hour and going by the shortest path. It would take us 32 days to reach the orbit of Jupiter. To cross the orbit of Saturn would take nearly two months, and it would be almost four months before we could hope to make our rendezvous with Uranus. If we decided to go on to the orbit of Neptune, we would have to prepare for another journey lasting for nearly 70 days. (Of course these figures are very rough, but they do serve to show that the outermost worlds are indeed remote.)

This may be the place to introduce Bode's Law, which gives a mathematical relationship between the distances of the planets from the Sun. It was first mentioned by an otherwise-obscure German astronomer named Titius, but was popularized in the 1770s by Johann Elert Bode. It may be summed up as follows:

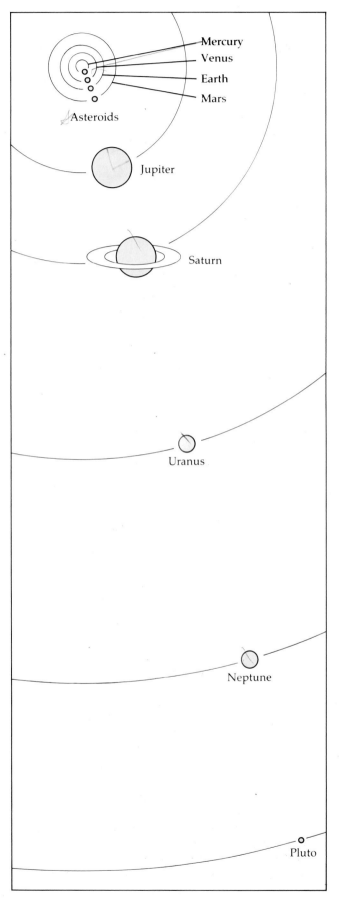

The orbits of the planets to scale.

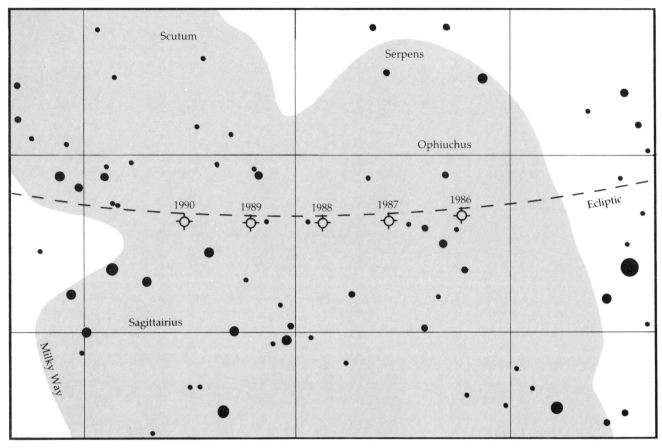

The position of Uranus on opposition dates 1986–1990.

Take the numbers 0, 3, 6, 12, 24, 48, 96, 192 and 384, each of which (apart from the first) is double its predecessor. Add 4 to each. Taking the Earth's distance from the Sun as 10, the distances of the other planets, out as far as Uranus (which was unknown when Bode's Law was first announced) are represented with tolerable accuracy (see Table 1).

Table 1 *Planetary distances*

Planet	Distance by Bode's Law	Actual distance on the Bode scale
Mercury	4	3.9
Venus	7	7.2
Earth	10	10.0
Mars	16	15.2
–	28	–
Jupiter	52	52.0
Saturn	100	95.4
Uranus	196	191.8
Neptune	388	–
Pluto	–	394.6

The 'missing number' 28 is accounted for by the asteroids; the first of them, Ceres, was discovered in 1801, and fits in well enough. When Uranus was discovered, in 1781, the agreement was regarded as confirmation that Bode's Law was of real significance. But it breaks down completely for Neptune – and 'manœuvring' to make the last Bode number agree with the distance of Pluto is not very convincing in view of Pluto's puny size and mass.

Quite apart from this, the agreement for even the inner planets is by no means perfect. All in all, it seems likely that Bode's Law is due to nothing more than coincidence, but more will be said about it later, because there are still a few astronomers who retain a degree of faith in it even though it is clearly non-valid at distances greater than that of Uranus.

As we have seen, the four giants have various features in common, but it would be misleading to suppose that they are similar in all respects. Nothing could be less true. Jupiter and Saturn make up one pair, though they are far from identical (quite apart from the glory of Saturn's ring-system). Uranus and Neptune make up another pair, and they are very similar in size and mass, but they are non-identical twins, and in composition and nature they differ strongly from either Jupiter or Saturn. Uranus is in many respects unusual. In particular, its axis of rotation in inclined at the extraordinary angle of 98 degrees to the perpendicular to the orbit, so that there are times – as at present – when one of the poles faces the Sun and is in constant daylight. The Uranian 'seasons' are peculiar by any standards.

Fifteen satellites are now known, as against only five before the flight of Voyager 2. All are smaller than our Moon (though, perhaps significantly, two of them are not much smaller than Pluto). There is always the chance that in the far future, manned flight to them may be achieved – so what would the view from the Uranian system be like?

As a vantage point for observing the rest of the Solar System, it would be very poor. The Sun would have an apparent diameter of less than two minutes of arc, though admittedly it would still be very brilliant; sunlight on one of the satellites of Uranus would be of the order of 1200 times that of full moonlight on Earth. Of the planets, Saturn would be fairly bright when well-placed, as would happen every 45.5 years – but remember, Saturn is on average slightly further away from Uranus than the Earth is from Saturn. Jupiter would reach a maximum magnitude of 1.7, but it would always be inconveniently close to the Sun in the sky, and would he hard to make out. Only Neptune would appear brighter than it does from Earth, though it would be barely visible with the naked eye even when at opposition.

Comets would, of course, by-pass Uranus, but at such tremendous distances from the Sun they would have no 'heads' and no 'tails'; all that would be left would be the nucleus, which, if Halley's Comet is any guide, is likely to be blacker than coal. The stars would appear just the same as they do from Earth, but otherwise an inhabitant of the Uranian system would be much less well-off than we are so far as observing is concerned.

Clearly we have learned a great deal, but let us bear in mind that Uranus has been known for only a little over two hundred years, and that it came to light because of the skill and dedication of one man – William Herschel.

William Herschel

Friedrich Wilhelm Herschel, possibly the greatest observational astronomer of all time, was born in Hanover on 15 November 1738. There was nothing about his background to indicate that he was destined for world fame. He was one of ten children, of whom four died young; his father was a bandmaster in the Hanoverian Guards, and the fact that young Wilhelm was well educated is surprising in view of the fact that his mother, Anna, never learned how to read and write. But at the local Garrison School he acquitted himself well. Languages, mathematics and music came easily to him, and by 1753 he was engaged as a musician in the Hanovarian Guards even though he was no more than fourteen years of age.

He was not destined to stay long in the Army, but it is not true to claim, as some people have done, that he deserted. The facts prove otherwise. The Seven Years' War was raging (at that time, remember, England and Hanover were united under one king, George II), and in 1756 the young musician was sent to England, with his regiment, to help guard against the possibility of a French invasion. He stayed there for some months – during which time he learned how to speak English – and then returned to Hanover to take part in active campaigning under the command of the Duke of Cumberland. In July 1757 the Duke's army was utterly routed at the Battle of Hastenbeck, and Herschel did admittedly take temporary and unauthorized 'leave of absence' after a night spent in a water-filled ditch, but after a few days he returned and applied for his discharge. At that stage of the war, musicians were not wanted. Moreover, Herschel had joined the Guards when he had been under age, so that he had never been formally enlisted. His release was granted – and a document signed by the local commander, General A.F. von Spörken, makes it clear that he left honourably. That document is now on view at the Herschel Museum in the city of Bath. So much for the story that he was an army deserter!

On his release, Herschel made for Hamburg. One of his brothers, Jacob, joined him, and the two made their way to London, arriving virtually penniless in late 1757. Apart from a few fleeting visits to the Continent, the rest of the story belongs to England, and henceforth the future astronomer was to be known as William instead of Wilhelm.

London, William realized, was 'over-stocked with musicians', and after a period during which he earned money by copying music he decided to try his luck further north. After a spell as a bandmaster in Yorkshire, he became a free-lance covering much of Northern England. In 1764 he paid a visit to his family in Hanover (Jacob had returned home some years earlier) and this was the last time that he saw his father. He also noted that his fourteen-year-old sister, Caroline, was in danger of becoming a household drudge, a point which was significant in view of what happened later.

His first really major appointment was as organist at Halifax, which he obtained after an audition which vastly impressed the maker of the organ, an elderly German-Swiss named Snetzler, who exclaimed that Herschel 'gave his pipes room to speak'. But by 1766 Herschel had travelled to Bath to become organist at the famous Octagonal Chapel. The new organ there – again built by Snetzler – was not ready, and until it was installed Herschel became an oboeist in Richard Linley's Orchestra which played daily in the Pump Room. Bath, of course, was then one of the greatest social centres of England, and William Herschel, with his good looks, pleasant personality and musical skill, fitted excellently into the scene.

On 4 October 1767 the Octagon Chapel was opened, with Herschel as organist, and his reputation spread. Several of his brothers visited him (one, Alexander, made his home in England), and in 1772 William

Sir William Herschel at the age of 56. (From a pastel by J. Russell RA 1794.)

brought over his sister Caroline both to keep house for him and to follow her own career as a singer. Had nothing else intervened, William would have remained a very successful and prosperous musician; he was appointed Director of Public Concerts in Bath, and everything went well.

Astronomy altered matters. William's first observations date back to 1766, when he looked at Venus and an eclipse of the Moon, but he did not take astronomy seriously, though he went so far as to borrow a small reflecting telescope which he found to be quite unsatisfactory. After moving to a new home – No. 7 New King Street, Bath – he bought some equipment from a neighbour, and decided to make his own reflector. Helped by Alexander, who was a skilful mechanic, and later by Caroline, he began to take astronomy very seriously indeed, and before long he was experimenting in a primitive 'telescope-making workshop'. Of course, all this had to be done in his limited spare time; his money came from his music.

19 New King Street, Bath: (a) exterior view; a commemorative plaque has been set up by the Herschel Society; (b) the drawing-room; (c) the dining-room; (d) the kitchen, in which Herschel carried out some of his experiments; (e) the furnace, built in the corner of the room during the mirror-making period. All these photographs are modern, taken in 1986 after the conversion of 19 New King Street into the present Herschel Museum.

c

d

e

a

b

Making a telescope mirror proved to be more difficult than he had expected; remember, too, that in those days glass could not be cast or worked with sufficient precision, so that telescope mirrors had to be made out of what is called speculum metal (an alloy of copper and tin). It is said that Herschel's first two hundred mirrors were failures, but he refused to give up, and at last he produced a reasonably good telescope with a mirror about five inches in diameter and a focal length of 5.5 feet. On 4 March 1772 he turned this instrument toward the Great Nebula in Orion, and his astronomical career may be said to have begun at that moment.

In the summer of 1774 they moved to another part of Bath (exactly where does not seem to be known), but three years later they came back to New King Street, this time to No. 19. Another move followed in 1779 to Rivers Street, a few doors away from the newly-formed Literary and Philosophical Society of Bath. The Rivers Street house had no garden, so that Herschel had to carry his telescopes outside to use them. By a

Herschel telescope, now in the Science Museum, South Kensington. Whether it is the actual telescope used for the discovery of Uranus is not certain, but it may well have been.

lucky chance, a passer-by saw him, and asked 'to have a look'. The newcomer was William Watson (later Sir William), who became a lifelong friend, and who communicated Herschel's first papers to the highly prestigious Royal Society.

Herschel had already learned a great deal about astronomy, and he decided to begin making 'reviews of the heavens', mainly in an attempt to find out the distances of the stars and their distribution in space. Rivers Street was not a good site – in March 1781 the Herschels moved back to No. 19 New King Street – and it was from here, on the thirteenth of the month, that William discovered the strange object which we now know to be the planet Uranus.

That moment changed his whole life. As soon as the nature of the new world was established, Herschel became internationally famous. The King, George III, was interested in astronomy, and even had a small observatory built in Kew; he gave Herschel the title of 'King's astronomer', and a pension which, though not over-generous, was enough to support him. Henceforth it was astronomy which became the profession, music the hobby.

The telescope used to find Uranus had a mirror 6.2 inches across, and a focal length of 7 feet (the dimensions and apertures of telescopes of this period are always given in Imperial measure, not metric). It was optically good, but Herschel wanted something much larger. He made a 12-inch, with a 20-foot focal length, which worked well, and he then began work on a 36-inch mirror. The results were somewhat hair-raising. A small furnace was built in the basement of No. 19, and all was prepared. Then, to quote from Caroline's diary:

The mirror was to be cast in a mould prepared from horse dung, of which an immense quantity was to be pounded in a mortar and sifted through a fine seaf . . .

Then, before the metal was fluid enough for casting, the mould cracked, and the fiercely-hot liquid spread across the floor.

Both my brothers and the caster and his men were obliged to run out at opposite doors, for the stone flooring (which ought to have been taken up) flew about in all directions as high as the ceiling. My poor brother fell, exhausted by heat and exertion, on a heap of brickbats.

That marked a temporary end to telescope-making, and at No. 19 New King Street it was never resumed, because it was made clear that Herschel's career would be greatly helped if he were closer to the Royal Court at Windsor. Reluctantly he packed up his belongings, sold No. 19, and moved first to a house in Datchet, then to Old Windsor, and finally to Observatory House, Slough.

Herschel was one of the most tireless of all observers. No clear night was wasted, and Caroline worked with him, acting as his assistant and recorder – even after William's decidedly unexpected marriage in 1788, after which Caroline moved out of Observatory House and stayed in a succession of lodgings close by. We must never forget Caroline's contribution. Without her, William would never have been able to carry out as much observational work as he actually did, and it is also worth noting that Caroline herself found time to discover several comets.

William's largest telescope, with a 49-inch mirror and a focal length of 40 feet, came into full use in August 1789, in the garden of Observatory House. It used the 'front-view' optical system; there was no secondary mirror, and the main mirror was tilted, producing an image which could be viewed direct through an eyepiece used by the observer in his 'cage' fixed to the main tube. Almost at once the light-grasp of the '40-foot' enabled Herschel to discover two new

Above *Model of the 'Uranus telescope' made by Michael Tabb, and now on display at 19 New King Street.*

Top right *Caroline Herschel at the age of 79, from the portrait in oils by Tielman in 1829.*

Bottom right *Monument to Herschel on the site where the 40-foot reflector was set up at Slough. It was then the garden of Observatory House, but today it is the courtyard of the Rank Xerox works.*

inner satellites of Saturn, those which we now call Mimas and Enceladus.[†] But it must be admitted, sadly, that the giant telescope was not really a success. It was clumsy and awkward to use; it did not perform even moderately well unless the air was absolutely calm and steady, which does not happen often; and the mirror tarnished easily, so that keeping it in tolerable condition was remarkably difficult. Most of Herschel's work was carried out with much smaller telescopes. But at least it was the largest of its time, and it was not surpassed until 1845, when the Earl of Rosse built his extraordinary but much more successful 'Leviathan' with a mirror no less than 72 inches across.

Herschel's activities spread to all fields of observation. He studied the elusive surface features of Venus, and was able to show that observations of high mountains there, reported by other observers, were spurious. He followed the markings on Mars, and gave a value for the rotation period of 24 hours 37 minutes 26.3 seconds, which is correct to within 4 seconds. He watched a transit of Mercury across the face of the Sun, and concluded, correctly, that the sharpness of the disk indicated the virtual lack of any Mercurian atmosphere. He studied comets, claiming that when near the Sun they emitted a certain amount of light on their own account, and also lost some of their material by evaporation. He spread out sunlight by means of a prism, and found that there were heating effects beyond the red or long-wave end of the visible band of colours, thereby laying the foundations of what we now call infra-red astronomy.

All in all, his greatest work was in connection with what he called 'the construction of the heavens'. He tried to measure star distances; and although he failed in this, he was led on to the discovery that many double stars are physically-associated or binary systems. By measuring the individual or proper motions of certain stars, he found that the Sun is moving through space toward a point in the constellation Hercules; he discovered thousands of new double stars, clusters and nebulae; and it was he who established that our star-system or Galaxy is a flattened system, so that the lovely band of the Milky Way is nothing more than a line of sight effect, caused when we look along the main plane of the system and see many stars almost one behind the other.

William Herschel received every scientific honour that the world could bestow, and in 1816 he was knighted (perhaps belatedly) by the Prince Regent. Of course he was not infallible, and some of his ideas sound strange today – notably his conviction that most worlds are inhabited, even the Sun! But his observational work was unrivalled, and it was fitting that in his last years he should become President of the new Astronomical Society of London, soon to become the Royal Astronomical Society. He died on 7 September 1822, active almost to the last. Caroline returned to Hanover, and lived on until 1848; William's son, John, followed in his father's footsteps, and became the first man to carry out really detailed observations of the stars of the far southern sky.

[†] Fittingly, the large crater on Mimas, discovered during the Voyager missions, has been named 'Herschel'.

The Milky Way according to Herschel, from his paper on the Construction of the Heavens.

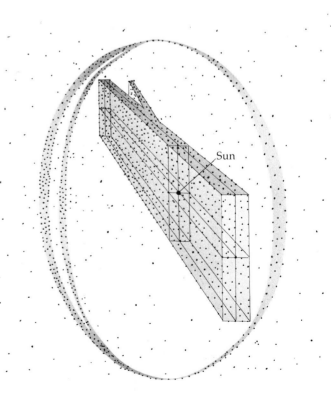

The Herschelian (front view) telescope.

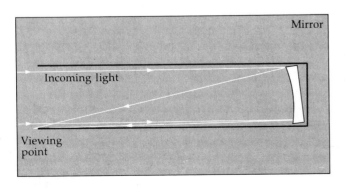

Observatory House was pulled down in 1960; parts of the tube of the 40-foot reflector are on view at the Old Royal Observatory in Greenwich Park, while the mirror is in the Science Museum, and the point in Slough where the great telescope used to stand has been marked by a sculpture in the courtyard of the Rank Xerox offices. But No. 19 New King Street has been preserved as a small Herschel Museum, and if you go there you can see many of the notebooks, drawings and instruments used by William and Caroline, together with a replica of the telescope with which Uranus was discovered. You can also go into the garden, and stand where we believe William to have stood when he made that great discovery over two hundred years ago.

Above Observatory House, Slough – one of the last photographs of it taken (by Patrick Moore) a day or so before demolition began.

Top right Remains of the tube of the 40-foot reflector, now on display at the Old Royal Observatory, Greenwich Park.

Bottom right Sir William Herschel's 40-foot telescope, which is also part of the crest of the Royal Astronomical Society.

The discovery of Uranus

On the night of 13 March 1781 William Herschel was alone at No. 19 New King Street, Bath. He and Caroline had been packing up from their previous home in Rivers Street, and Caroline had stayed behind to complete some business matters; but William, as always, was anxious not to lose a single night's observing, and the skies were clear. He had set up his 6.2-inch mirror, 7-foot focus reflector, and had begun work on his 'review of the heavens'. He was surveying star-field after star-field, counting the stars in them and noting any unusual objects; his aim was to find out the form of the Galaxy, and any special discoveries which he happened to make were sheer bonus.

We have to admit that we are not quite sure which telescope Herschel was using on that particular night, because he made several which were to all intents and purposes identical. It may have been the one now on display at Greenwich. Certainly the replica at No. 19 New King Street is identical in every respect. It was a typical Herschel instrument, with its wooden tube; of course it had an altazimuth mounting – equatorials still lay in the future.

Herschel was busy examining an area in the constellation of Gemini, the Twins, when he came across something which caught his attention. In his own words:

> On Tuesday the 13th March, between ten and eleven in the evening, while I was observing the small stars in the neighbourhood of H Geminorum, I perceived one that appeared visibly larger than the rest; being struck with its uncommon magnitude, I compared it to H Geminorum and the small star in the quartile between Auriga and Gemini, and, finding it so much larger than either of them, suspected it to be a comet.
>
> I was then engaged in a series of observations on the parallax of the fixed stars which I hope soon to have the honour of laying before the Royal Society; and these observations requiring very high powers I had ready on hand the several magnifiers of 227, 460, 932, 1536, 2010 etc., all of which I successfully used on that occasion. The power I had on when I first saw the comet was 227. From experience I knew that the diameters of the fixed stars are not proportionally magnified with higher power as the planets are; I therefore put on the powers 460 and 923 and found the diameter of the comet increased in proportion to the power, as it ought to be on the supposition of its not being a fixed star, while the diameters of the stars with which I compared it were not increased in the same ratio. Moreover the comet, being magnified much beyond what its light would admit of, appeared hazy and ill-defined with these great powers, while the stars preserved that lustre and distinctness which from many thousand observations I knew they would retain. The sequal has shown that my surmises were well founded, this proving to be the comet we have lately observed.'

This extract comes from a paper read to the Royal Society on 26 April (communicated by the faithful Watson; Herschel had not then been made a Fellow). But was the object really a comet, as Herschel originally, and naturally, believed? No thought of a new planet was in anybody's mind, least of all Herschel's. But already the movements of the 'comet', as followed not only by Herschel but also by various other observers, were making it highly suspect. For one thing it was very gradual in its motions. At an early stage – we are not sure when – Herschel had notified the famous French comet-hunter Charles Messier, and Messier had replied expressing his astonishment at Herschel's discovery simply because the shift was too slight to be noticed over a period of only 24 hours. Herschel replied that he had found the object to be non-stellar purely because it had not looked like a star; it was not for another night or two that he was able to measure its motion. (Messier, of course, habitually used low magnifications for his comet-hunts, while Herschel preferred magnifications so high that even his contemporaries were sometimes sceptical about them.)

The Astronomer Royal of the time was the Rev. Nevil Maskelyne, founder of the *Nautical Almanac*. Herschel very properly informed Maskelyne about the discovery of the 'comet', and he also contacted the Rev. T. Hornsby at Oxford, since at that time Greenwich and Oxford were the only professional observatories in England. The position of the object was definite enough; Herschel's 'H Geminorum' was the star we now call 1 Geminorum, of magnitude 4.3, while 'the small star in the quartile' is 132 Tauri, of magnitude 5. Neither was the 'comet' faint; it was on the fringe of naked-eye visibility. On 19 March, less than a week after the discovery, Herschel recorded that 'The Comet's apparent motion is at present 2¼ seconds of arc per hour'. On 25 March: 'The apparent motion is accelerating, and its apparent diameter seems to be increasing.' On 6 April: 'With a magnifying power of 278 times the Comet appeared perfectly sharp upon the edges, and extremely well defined, without the least appearance of any beard or tail.'

Hornsby, at Oxford, was unable to identify the object for several weeks, and wrote to Herschel for further information. Herschel duly provided it, and on 14 April Hornsby finally located the object, after which he realized that he had already 'unknowingly' seen it on 29 and 30 March. He wrote to Herschel again: 'I do not in the least question but this is the comet of 1770, but whether it has passed its Perihelion or has not yet come to it, is more than I can say at present. I will very soon try to construct its orbit.'

The reference to the 1770 comet is interesting. In fact this particular comet – discovered by Messier, but always remembered as Lexell's Comet from the mathematician who computed its path – had come very close to the Earth, passing by us as only around

2.5 million kilometres, which remains a record. It had been an easy naked-eye object, and much later, in 1779, Anders Lexell, a Finnish mathematician working in St. Petersburg, had worked out a period of 5.6 years. This would have meant a return in 1781; that of 1776 had, said Lexell, been missed because the comet had been so badly placed. Actually, we know that the comet had a close encounter with Jupiter between 1776 and 1781, and the orbit was violently perturbed. The current period must be over 250 years, perhaps more, and we have no chance of recovering Lexell's Comet in the future except by sheer luck. However, Hornsby did not know this, and his suggestion was a very reasonable one.

Nevil Maskelyne was not so sure. Even before Herschel's classic paper had been read at the Royal Society meeting, Maskelyne had written to him: 'I am to acknowledge my obligation to you for the communication of your discovery of the present Comet or planet, I don't know which to call it. It is as likely to be

Herschel's paper read to the Royal Society on 26 April 1781.

XXXII. *Account of a Comet. By Mr.* Herschel, *F. R. S.; communicated by Dr.* Watson, *Jun. of* Bath, *F. R. S.*

Read April 26, 1781.

ON Tuesday the 13th of March, between ten and eleven in the evening, while I was examining the small stars in the neighbourhood of H Geminorum, I perceived one that appeared visibly larger than the rest: being struck with its uncommon magnitude, I compared it to H Geminorum and the small star in the quartile between Auriga and Gemini, and finding it so much larger than either of them, suspected it to be a comet.

I was then engaged in a series of observations on the parallax of the fixed stars, which I hope soon to have the honour of laying before the Royal Society; and those observations requiring very high powers, I had ready at hand the several magnifiers of 227, 460, 932, 1536, 2010, &c. all which I have successfully used upon that occasion. The power I had on when I first saw the comet was 227. From experience I knew that the diameters of the fixed stars are not proportionally magnified with higher powers, as the planets are; therefore I now put on the powers of 460 and 932, and found the diameter of the comet increased in proportion to the power, as it ought to be, on a supposition of its not being a fixed star, while the diameters of the stars to which I compared it were not increased

in

in the same ratio. Moreover, the comet being magnified much beyond what its light would admit of, appeared hazy and ill-defined with these great powers, while the stars preserved that lustre and distinctness which from many thousand observations I knew they would retain. The sequel has shewn that my surmises were well founded, this proving to be the Comet we have lately observed.

I have reduced all my observations upon this Comet to the following tables. The first contains the measures of the gradual increase of the Comet's diameter. The micrometers I used, when every circumstance is favourable, will measure extremely small angles, such as do not exceed a few seconds, true to 6, 8, or 10 thirds at most; and in the worst situations true to 20 or 30 thirds: I have therefore given the measures of the Comet's diameter in seconds and thirds. And the parts of my micrometer being thus reduced, I have also given all the rest of the measures in the same manner; though in large distances, such as one, two, or three minutes, so great an exactness, for several reasons, is not pretended to.

TABLE

TABLE I. Measures of the Comet's diameter *.

Days.	" '''	Powers.
March 17	2 53	932. 460.
19	2 59	932. 460.
21	3 38	460.
28	4 7	932 } these measures agree to 9'''.
—	3 58	227
29	4 7	227 rather too small a measure.
—	4 25	227 seems right.
April 2	4 25	227
6	4 53	227
15	5 11	227 very good; not liable to half a second of error.
—	5 20	227
18	5 2	227 true to 12 " or 18 " at most.

Having measured the diameter of the Comet with such high power as 932 and 460, it may not be amiss to make one observation on this subject, lest it should be misapprehended that I pretend to a distinct power of such magnitude upon all celestial objects in general. By experience I have found, that the aberration or indistinctness occasioned by magnifying much, provided the object be still left sufficiently distinct, is rather to be put up with, than the power to be reduced, when the angles to be measured are extremely small. The reason of this may, perhaps, be that a small error of judgement, to which we are always liable, is of great consequence with a low power, as bearing a considerable proportion to the diameter of the object;

* There are several optical deceptions which may affect the measures of objects that subtend extremely small angles. Thus I have found, by experience, that a very small object will appear something less in a telescope when we see it first than when we become familiar with it. There is also a deflection of light upon the wires when they are nearly shut; but as none of these deceptions are well enough understood to apply a correction, I leave them affected with them.

whereas

whereas with a higher power the proportion of this error to the whole becomes much lefs, and the meafure more exact, even after we have made allowance for a fmall additional error occafioned by the want of that perfect diftinctnefs which is required for other purpofes. However, to enter deeply into an explanation of this would lead me to fpeak of the caufes of the aberration of rays in the focus of an object fpeculum, of which there are fome that are feldom taken into confideration by opticians, and indeed are fuch as cannot be calculated ; but this not being my prefent purpofe, fuffice it to obferve, that the method is juftified by experience.

When the diameter of the Comet was increafed to about 4″, I thought it advifable to leffen the power with which I meafured ; and, as I made ufe of two different micrometers, as well as eye-glaffes, I took a meafure with both of them. The agreement of the micrometers to 9‴ is no fmall proof of the goodnefs of the obfervations of the 28th of March, and very properly connects the meafures of the high powers with thofe that were made with 227.

TABLE II. Diftance of the Comet from certain telefcopic fixed* ftars which I have marked α, β, γ, δ, ε, ζ.

D. H. M.		° ′ ″ ‴	
Mar.13 10 30	from α, fig. 1*.	2 48 0	by pretty exact eftimation true to 20″.
17 11 0		0 41 58	by the micrometer and power 227.
18 7 20		1 0 35	
— 9 16		1 6 59	
— 10 55		1 10 40	
19 7 4		1 46 40	
— 10 42		1 51 23	
21 10 0		3 39 46	
24 8 12	from β, fig. 2.	2 55 39	true to 4 or 5″, an indifferent obfervation.
— 10 58		2 53 4	true to 4 or 5″.
25 7 24		2 12 46	true to 2 or 3″.
— 9 47		2 14 18	
26 10 43		1 48 3	true to 2 or 3″.
28 7 46		2 55 49	true to 4 or 5″.
29 8 50	from γ, fig. 3.	2 20 51	true to 2″.
30 7 55		1 28 48	true to 2 or 3″.
Apr. 1 7 45		2 39 20	
6 8 50	from λ, fig. 4.	2 51 23	
15 10 18	from ι, fig. 5.	4 27 57	eftimated by the field, true to 5 or 6″.
16 7 50		3 9 14	by the micrometer, true to 3 or 4″.
— 10 47		2 50 56	true to 3 or 4″.
18 8 18		3 18 4 }	mean 3′ 17″, true to 1″ or 1¼.
		3 15 57 }	
— 8 50	from ζ, fig. 6.	2 24 57	
19 8 38		3 2 5	true to 3 or 4″.

* The figures are drawn upon a fcale of 80 feconds to one inch.

TABLE

TABLE III. Angle of pofition of the Comet with regard to the parallel of declination of the fame telefcopic fixed ftars meafured by a micrometer, of which I have given the defcription, and a magnifying power of 278. See fig. 1. 2. 3. 4. 5. 6.

D. H. M.		° ′	
Mar.13 10 30	B α Comet,	0 0 }	by fuperficial eftimation, liable to an error of 10 or 12 degrees.
17 11 0	A α Comet, fig. 1*.	89 56	by the micrometer.
18 8 20		56 39	
— 9 24		41 33	true to 1°.
19 7 23		29 47	true to 1°.
21 10 10		11 46	true to 4 or 5°.
— 11 48		12 14	
24 8 23	B β Comet, fig. 2.	38 39	true to 2 or 3°.
— 11 4		36 14	true to 3 or 4°, air very tremulous.
25 7 33		53 18	
— 9 55		56 32	liable to a confiderable error.
26 10 55	A β Comet,	87 0	true to 2 or 3°.
28 7 58		28 51	true to 3 or 4°.
29 9 25	B γ Comet, fig. 3.	32 19	true to 1 or 2°.
30 8 25		72 14	true to 3 or 4°.
Apr. 1 7 55	A γ Comet,	28 51 well taken, }	27° 46′, true to 1°.
—		27 14 more exact, }	
6 8 28	B δ Comet, fig. 4.	84 42	true to lefs than 2°
15 10 27	B ι Comet, fig. 5.	29 9	true to 2 or 3°.
16 8 1		49 11	true to 1°.
— 10 55		50 47	true to 1½ or 2°½.
18 8 31	A ι Comet,	47 9 very well taken, }	47°, true to lefs than1°.
—		46 35 pretty well, }	
— 9 8	B ζ Comet,	82 39	
19 8 56	A ζ Comet, fig. 6.	48 18 }	49° 3′, true to 1°.
—		49 48 }	
— 10 45		47 30	true to 2 or 3°.

* The angles are drawn true to the meafure, without allowing for errors.

Mifcellaneous obfervations and remarks.

March 19. The Comet's apparent motion is at prefent 2½ feconds *per* hour. It moves according to the order of the figns, and its orbit declines but very little from the ecliptic.

March 25. The apparent motion of the Comet is accelerating, and its apparent diameter feems to be increafing.

March 28. The diameter is certainly increafed, from which we may conclude that the Comet approaches to us.

April 2. This evening at 8 h. 15′ the Comet was a little above the line drawn from η to θ in fig. 7. This figure is only delineated by the eve, fo that no very great exactnefs in the diftances of the ftars is to be expected; but I fhall take the firft opportunity of meafuring their refpective fituations by the micrometer.

April 6. With a magnifying power of 278 times the Comet appeared perfectly fharp upon the edges, and extremely well defined, without the leaft appearance of any beard or tail.

April 16. Fig. 8. reprefents the fituation of the Comet this evening about nine o'clock, and is only an eye-draught of the telefcopic ftars.

Remarks on the path of the Comet.

We may obferve, that the method of tracing out the path of a celeftial body by taking its diftance from certain ftars, and the angle of pofition with regard to them, cannot be expected to give us a compleatly juft reprefentation of the tract it defcribes, fince even the moft careful obfervations are liable to little errors, both from the remaining imperfections of inftruments, though
they

they fhould be the moft accurate that can be had, and from the difficulty of taking angles and pofitions of objects in motion. Add to this a third caufe of error, namely, the obfcurity of very fmall telefcopic ftars that will not permit the field of view fo well to be enlightened as we could wifh, in order to fee the threads of the micrometer perfectly diftinct.

This will account for the apparent diftortions to be obferved in my figures of the Comet's path. Some little irregularity therein may alfo proceed from different refractions, as they have not been taken into account, though the obfervations have been made at very different altitudes, where confequently the refractions muft have been very different. But though this method may be liable to great inconveniences, the principal of which is, that many parts of the heavens are not fufficiently ftored with fmall ftars to give us an opportunity to meafure from them, yet the advantages are not lefs remarkable. Thus we fee that it enabled me to diftinguifh the quantity and direction of the motion of this Comet in a fingle day (from the 18th to the 19th of March) to a much greater degree of exactnefs than could have been done in fo fhort a time by a fector or tranfit inftrument; nay even an hour or two, we fee, were intervals long enough to fhew that it was a moving body, and confequently, had its fize not pointed it out as a Comet, the change of place, though fo trifling as 2¼ feconds *per* hour, would have been fufficient to occafion the difcovery. A gentleman very well known for his remarkable fuccefs in detecting Comets * feems to be well aware of the difficulty to difcover a motion in a heavenly body by the common methods when it is fo very fmall; for in a letter he favoured me with, fpeaking of the Comet, he fays: " Rien n'etoit plus difficile que de la " reconnoître et je ne puis pas concevoir comment vous avés pu

* Monf. MESSIER.

" revenir

" revenir plufieurs fois fur cette étoile ou Comête; car abfolu- " ment il a fallu l'obferver plufieurs jours de fuite pour s'ap- " perçevoir qu'elle avoit un mouvement."

I need not fay that I merely point this out as a temporary advantage in the method I have taken; for as foon as we can have regular, conftant, and long continued obfervations by fixed inftruments, the excellence of them is too well known to fay any thing upon that fubject: for which reafon I failed not to give immediate notice of this moving ftar, and was happy to furrender it to the care of the Aftronomer Royal and others, as foon as I found they had begun their obfervations upon it.

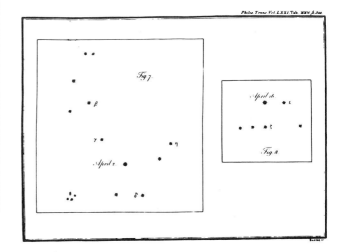

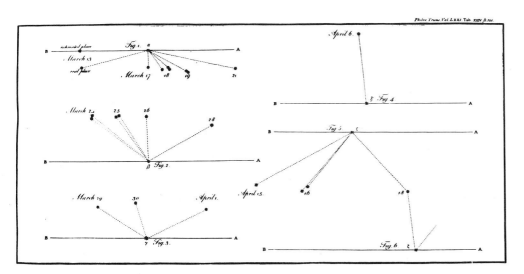

a regular planet moving in an orbit nearly circular to the Sun as a Comet moving in a very eccentric ellipsis. I have not yet seen any coma or tail to it.' Neither had Herschel – and if Herschel could not do so, then certainly nobody else could, because nobody else had his observational skill and nobody else would dare to use such high magnifications on a reflector with a mirror only just over 6 inches aperture.

It did not take long for the nature of the object to be cleared up. There were only two possibilities – comet or planet – and the orbit would show which must be true. Various mathematicians set to work, using all the observations they could secure, including those by Herschel, Hornsby and Messier. Among the mathematicians were Ruggieri Boscovich of Rome, whose reputation was very high, and the French aristocrat

Bouchart de Saron, friend of Messier, who was later guillotined during the Terror – one of the most tragic and unjustified scientific victims of the Revolution. But it seems that the first 'circular' orbit, proving that the object must be a planet, was produced by Anders Lexell. Lexell gave the period as 82 years 10 months, only a little lower than the real value, and the mean distance from the Sun as 2840 million kilometres, compared with the correct value of 2870 million kilometres. This was a remarkably good result, and it is surprising that as late as February 1782 Hornsby could still write to Herschel that 'it is the fashion I think now to call it a new star or planet, but I cannot help thinking it will prove to be a comet'.

By August 1782 Herschel was installed at his new home near Windsor, thanks to his new status as Astronomer to George III (note, by the way, that this was an unique title which has never been repeated; during Herschel's career the Astronomers Royal were Maskelyne, until 1811, and then the Rev. John Pond). Maskelyne's letter gave unqualified congratulations to Herschel for what he had achieved, and ended: 'I hope you will do the astronomical world the favour to give a name to your planet, which is entirely your own, and which we are so much obliged to you for the discovery of.'

Unwittingly, Maskelyne had sparked off a controversy which was not finally resolved until after Herschel's death. The naked-eye planets had been names after the Olympian gods, though the Roman names are always used instead of the Greek (Mercury instead of Hermes, Jupiter instead of Zeus, and so on). In the year of the discovery of the new planet, two names were proposed. Johann Elert Bode, of Bode's Law fame, suggested 'Uranus', because in mythology Uranus was the father of Saturn and therefore the first supreme Olympian; several eminent astronomers supported the choice. Alternatively, Jean Bernouilli, of the Berlin Academy, proposed 'Hypercronius' i.e. 'above Saturn'.

William Herschel had a different idea. In the summer of 1782 he wrote to his friend Watson suggesting that in honour of King George III, the planet should be named 'Novum sidus Georgianum'. Watson, a better Classical scholar, modified this to 'Georgium Sidus', and this was the name that Herschel always used for it; others referred to it as 'the Georgian Planet' or simply 'the Georgian'. Other nations were not in the least enthusiastic, which is hardly surprising, and in France another eminent astronomer, J. J. F. Lalande, suggested that the planet should be called 'Herschel'. Lexell went back to mythology and proposed 'Neptune', in honour of the sea-god son of Jupiter – though he added, probably in light-hearted vein, that this might be extended as 'Neptune de George III' or 'Neptune de la Grande Bretagne'. Yet another choice was 'Cybele', in mythology one of Saturn's wives.

Several of these rival names were in use for a curiously long time. 'Herschel' was still used by the French until well into the 1840s, even after the discovery, in 1846, of the planet which was soon christened 'Neptune'. ('Cybele', in fact, has been allotted to an asteroid, No. 65 of the swarm.) But the tardiest recognotion of all was in the *Nautical Almanac*. Only after 1850 did the Almanac replace 'the Georgian' with Bode's choice of 'Uranus'.

There are also discussions about how the name should be pronounced. There are two main variants: 'You-*ray*-nus' and '*You*-ranus'. Astronomers in general use the second. At least there is no argument about the symbol for the planet: ♅ , incorporating the first letter of Herschel's surname.

On all counts the discovery of Uranus must rank as one of the milestones in modern astronomy. For the first time in recorded history a new planet had been added to the Solar System, and the credit was due to a hitherto-unknown and completely unqualified amateur. Though Herschel went on to become world-famous, and though his work on 'the construction of the heavens' was in the end his most important contribution to science, it is for his identification of Uranus, from the garden of No. 19 New King Street in Bath, that he will always be best remembered. Both the present authors were at Mission Control on that day in January 26 when Voyager 2 made its closest approach to Herschel's planet, and it is surely fitting to reproduce here the message which was sent, and read out at the Conference, from the then Mayor of Bath:

On the occasion of the close approach of the Voyager 2 space-craft to the planet Uranus may I, on behalf of the citizens of Bath, England, and of the William Herschel Society, send a message of hope and congratulations to all those involved in this great achievement . . . Here in Bath we feel that now, almost 205 years after the discovery, information from the planet will become available which will vastly extend the work commenced over here two centuries ago by one of the most distinguished astronomers who ever lived. The story of his epoch-making discovery is recorded in his report to the Royal Society of 26 April 1781. In this Sir William described how he first suspected his observation to be that of a comet . . . At 19 New King Street, Bath, therefore, began the journey of knowledge which has culminated in today's climactic and triumphant achievement in the outer reaches of the Solar System. It is fitting that this message of greeting and goodwill to those who have made it possible should come from Bath, where the planet was first discovered, to the location of the most recent and sophisticated accomplishment stemming from that discovery, in Pasadena, California. We honour your achievements alongside those of Sir William Herschel.

Pre-discovery observations

Seven is the mystical number of the ancients, and so it had seemed only right that there should be seven members of the Solar System apart from the Earth: the Sun, the Moon, and the five naked-eye planets. The discovery of a new planet was quite unexpected. Admittedly, there is an old Burmese tradition which states that there are eight planets – the extra one being Ráhu, which is 'invisible', but it does not seem at all likely that Ráhu can have been Uranus.

Yet since Uranus, at its best, is dimly visible with the naked eye, we are entitled to ask: 'Why was it not seen before 1781?' The answer is that it was – at least 22 times, and no doubt many more. But it had always been mistaken for a star, and the fact that Herschel singled it out at first sight is due to the high magnification he was using on his telescope.

This, incidentally, may well be why even Herschel, during his 'reviews of the heavens', failed to discover any of the asteroids, even though one of them (Vesta) may become almost as bright as Uranus and can be glimpsed with the naked eye, while Ceres, Pallas and Iris all have mean opposition magnitudes of above 8. Neptune, outermost of the giant planets, has a mean opposition magnitude of 7.7. But to Herschel all the asteroids, plus Neptune, would have looked virtually stellar; he would have had to identify them by means of their movements from one night to another, and it is understandable that he did not succeed, even though he was able to make useful positional observations of the first four known asteroids once they had been found.

When Uranus was discovered, Johann Elert Bode set to work to see whether he could locate any early observations which might be useful in working out the orbit of the new planet. The greatest observer of pre-telescopic times was the eccentric Dane, Tycho Brahe, who compiled a magnificent star catalogue from his home at Hven, an island in the Baltic, between 1576 and 1596, and whose work enabled Johannes Kepler to show that the Sun rather than the Earth is the centre of the Solar System. Bode examined Tycho's catalogue, and came to the conclusion that a star recorded in the constellation of Capricornus might have been Uranus. However, when better orbital elements for Uranus had been computed, Bode realized that Tycho's observation did not fit.

What about the star catalogue compiled by the first Astronomer Royal, the Rev. John Flamsteed, with the aid of telescopic sights? Greenwich Observatory was founded in 1675, by express order of King Charles II, purely for navigational purposes. British seamen frequently lost their way when out of sight of land, mainly because although they could find their latitude easily enough (by measuring the altitude of the Pole Star above the horizon) it was a difficult matter to find longitude. The best method was to use the shifting Moon as a kind of 'clock-hand' against the stars, but this involved using a very accurate star catalogue, and Tycho's was not good enough. So the Royal Observatory came into being, and eventually Flamsteed produced an excellent catalogue.

Flamsteed decided to take each constellation and allot its stars definite numbers, in order of right ascension. Looking through the catalogue many years later, Bode came across a star, 34 Tauri, which did not appear to exist at all. It was near the star Omega Tauri in the Bull; Flamsteed had recorded it on 23 December 1690, and had given its magnitude as 6, just about on the limit of naked-eye visibility.

Where was 34 Tauri? At one stage Herschel himself, following a letter from Bode, believed that he had seen it, but further checks showed that there was nothing there. Careful mathematical calculations proved that the position of Flamsteed's star, as given on 23 December 1690, really did agree with that of Uranus, and the identification does not seem to be in doubt. Then, from the catalogues, it emerged that Flamsteed had recorded Uranus on five other occasions, once in 1712 and four times in 1715, when it had moved into the constellation of Leo (the Lion). Always it had been mistaken for a star.

Uranus was also seen three times by the third Astronomer Royal, James Bradley, in 1748, 1750 and 1753. Tobias Mayer recorded it in 1756, by which time it had shifted into Aquarius (the Water-bearer). But the record stands to the credit – or otherwise! – of a French observer, Pierre Charles Le Monnier, whose connection with Uranus is indeed curious.

Le Monnier himself was an unusual character. He became Astronomer and Professor of Physics in Paris, and carried out some excellent work, but he was not exactly popular, and it has been said that he never failed to quarrel with anyone whom he met. Neither was he as methodical as might have been expected, and it was related by one later critic that an observation which Le Monnier had made of Uranus was found scrawled upon a paper bag which had once contained hair perfume. Whether this story is true or not is uncertain – very probably it is false – but we do know that Le Monnier recorded Uranus at least ten times between 1764 and 1771, including a run of six observations in January 1769. The planet was then in Aries, one of the more barren constellations of the Zodiac, so that there should have been no problem in detecting an object which moved.

It is easy to laugh at Le Monnier; he had a major discovery within his grasp and missed it. But it has since been pointed out that in January 1769 Uranus was near its 'stationary point'; it is at best a slow mover, and would not be easy to identify under such conditions. Therefore we may exonerate Le Monnier to a considerable extent, but it is undoubtedly true that if he had been disposed to check his observations he could have identified Uranus more than a decade before Herschel did so. Le Monnier died in 1799; by then he had realized that he had made several pre-discovery observations.

As soon as Uranus had been identified, it was kept under constant observation, and its orbit was worked out. It was found to lie at a distance in reasonably good agreement with Bode's Law, and by 1820 Pierre Simon de Laplace had given a revolution period of 84.02 years, which is virtually correct (the modern value is 84.01 years). It was not for some time that any discrepancies began to show up – though when they did, they caused a great deal of perplexity.

Slowly but surely, Uranus began to wander from its predicted path. Table's of its motion, based upon all the observations available, did not explain how the planet moved. Finally Alexis Bouvard, a French mathematician of great skill and renown, decided that there must be serious errors in the pre-discovery observations which he had used together with those made since 1781. Therefore he discarded all the pre-discovery observations, and in 1821 produced a new set of tables based only upon the measurements which he knew to be reliable. When these also failed to explain the wanderings of Uranus, it became evident that there was some error of unknown nature. As we now know, the fact that Uranus failed to keep to its predicted path was because of the pull of the more remote planet Neptune, and it was from these clues that Neptune was tracked down on 1846.

Can we hope to find any observations of Uranus going back further than that recorded by Flamsteed on 23 December 1690? Probably not; even if we could, they would be of no more than historical value. So far as we can tell, therefore, the first definite observation of Uranus was made less than three hundred years before Voyager 2 flew past the planet and sent us back pictures from close range.

The position of Flamsteed's 34 Tauri (Uranus) 1690.

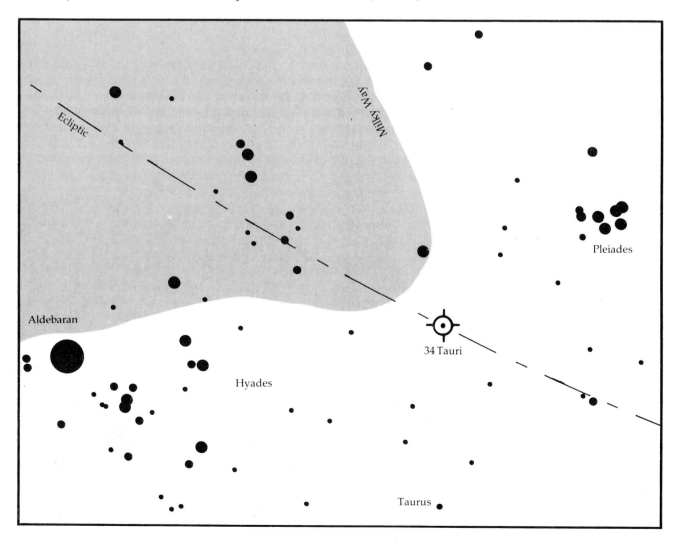

Early theories of Uranus

As soon as Uranus had been identified as a planet, efforts were made to see detail upon its surface. After all, most of the naked-eye planets show features when observed with sufficient telescopic power, though Venus does often appear blank, while details on Mercury are excessively hard to see even with powerful instruments (practically all our knowledge of the Mercurian surface has been derived from one spaceprobe, Mariner 10). Mars, of course, shows obvious patches, from which Herschel was able to measure the rotation period; there are also the white polar caps. Jupiter's belts are evident with modest optical power, while the disk of Saturn also shows dusky belts and bright zones which are distinct even though they are so often neglected in favour of the superb ring-system.

Uranus was clearly greenish in colour, but its small apparent diameter (never more than 3.7 seconds of arc) means that it has to be observed with a very high magnification to give the slightest hope of seeing anything on it. Herschel did his best, but with no success. He believed the disk to be flattened, in the same way as those of Jupiter and Saturn, and he measured the equatorial diameter as being 55 067 kilometres, which we now know to be slightly too great. But a discovery of quite a different kind led him to uncover one of the greatest mysteries of Uranus – the inclination of the axis.

In January 1787 he made an alteration to the optical system of his '20-foot' reflector. He removed the flat secondary mirror, and tilted the main speculum, so that the image of the object under study could be seen direct through the eyepiece from the cage or platform fixed to one side of the upper end of the tube. This is always known as Herschelian or front-view arrangement. It sounds excellent, because it removes the need for a second mirror, and each time a ray of light is reflected it loses a little of its brightness. Unfortunately there are major disadvantages too, and by now the system has fallen into disuse, but Herschel was very satisfied with it during his early experiments. 'I wondered why it had not struck me sooner,' he recorded. And before long he was able to announce that Uranus is attended by satellites.

By 28 February 1794 he claimed that altogether he had found no less than six satellites. In 1801 he observed another. But for once he erred. Of his supposed satellites, the two discovered in 1787 are genuine; we call them Titania and Oberon. Another may also have been genuine; the satellite found in 1802, Umbriel – but even if so it remained unconfirmed, simply because nobody except Herschel had the optical power of the observational skill to see it. Umbriel was definitely seen in 1851 by the English amateur William Lassell, together with an inner satellite, Ariel, which Herschel never saw. There is a suggestion that one of the inner satellites may have

been glimpsed by the Russian astronomer Struve, at Dorpat Observatory in Estonia, in 1847, but evidence is inconclusive.

Herschel is always credited with the discoveries of Titania and Oberon, whether or not he detected Umbriel is uncertain. Meanwhile it may be interesting to give (see Table 2) the 'satellite system' as it was believed to be as recently as 1860.

Table 2 *The satellites of Uranus in 1860*

Satellite	'Discovery'	Mean distance from Uranus, kilometres	Period, days	Notes
1	1851: Lassell	206 499	2.52	Ariel
2	1802: Herschel	287 821	4.14	Umbriel?
3	1790: Herschel	364 470	5.89	(Non-existent)
4	1787: Herschel	472 116	8.71	Titania
5	1794: Herschel	550 940	10.98	(Non-existent)
6	1787: Herschel	631 543	13.46	Oberon
7	1790: Herschel	1 263 141	38.08	(Non-existent)
8	1794: Herschel	2 526 004	107.69	(Non-existent)

This table has been taken from the *Handbook of Astronomy* by G.F. Chambers (Oxford University Press, Oxford 1861 edition) but it sums up the accepted version of the satellite system as it was then believed to be. In the 1889 edition of the book it is accepted that the extra satellites reported by Herschel do not exist.

Satellites of Uranus known up to 1860.

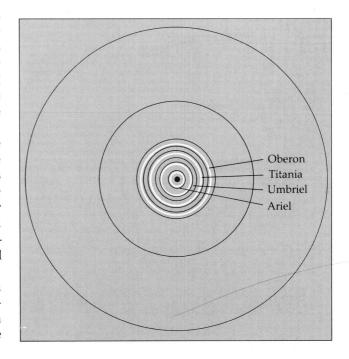

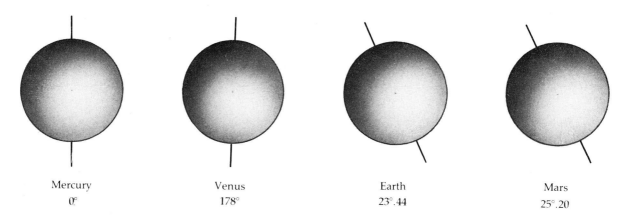

Mercury	Venus	Earth	Mars
0°	178°	23°.44	25°.20

The periods and distances of the real satellites as given in this table differ somewhat from the modern values. Thus the mean distance of Titania from Uranus is only 436 300 kilometres. But, so far as Titania and Oberon were concerned, the identifications were positive from the outset, and Herschel was able to follow their orbital movements. From this, he found that by assuming that they moved in the planet's equatorial plane, or nearly so, then the axis of Uranus must be inclined at something like a right angle!

This was a remarkable result. The axial inclinations of Mars and Saturn were known to be not very different from that of the Earth, and that of Jupiter considerably less (there was no information then about Mercury or Venus). Therefore, the proposed tilt of Uranus was totally unexpected, but the only alternative was to suppose that the orbits of the satellites were more polar than equatorial, which seemed even less likely.

More than a century after Lassell recorded Umbriel and found Ariel, G. P. Kuiper detected an inner satellite, Miranda, which Voyager 2 has shown to be one of the most complex objects in the Solar System. Voyager 2 also detected another ten small satellites; but all these are closer-in than Miranda, and they are not identical with those reported by Herschel. There seems, in fact, no doubt that Satellites 3, 5, 7 and 8 in the old table were either faint stars, or else 'optical ghosts', due to the front-view optical arrangement which Herschel was using.

It was also the front-view system which led to another illusion – the reported discovery of a ring.

Using a high magnification on his 20-foot reflector on 4 March 1787, Herschel found four projecting points of light, almost at right angles to each other, which he thought might be 'a double ring', that is two rings at right angles to each other. The effect was again seen on 7 March, but on the following night he came to

the conclusion that only one ring existed. He drew it again on 22 February 1789 ('The ring is short, not like that of Saturn, and this may account for the great difficulty of verifying it. It is descernible that the two ansae seem of a colour a little inclined to red') and on 26 February 1792 ('My telescope is extremely distinct; and when I adjust it upon a very minute double star, which is not far from the planet, I see a very faint ray like a ring crossing the planet, over the centre'). But already Herschel was becoming suspicious; the so-called ring could, he thought, be due to a slight scratch on his telescope mirror, or some other optical effect. His last mention of the ring is entered in his notebook for 4 December 1793. Finally he seems to have realized that his ring did not exist.

In this, of course, he was right. Later observations showed no trace ot it, and it was firmly relegated to the realm of myth. The true ring-system was not discovered until 1977, more or less fortuitously. There is absolutely no chance that Herschel could have seen it; indeed, any sign of it taxes the power of our most modern telescopes with our most modern electronic equipment.

Details on the disk itself were lacking. It was reasonable to assume that at so tremendous a distance from the Sun, the temperature must be very low, but even this was not regarded as certain. In 1838 Thomas Dick, a well-known astronomical writer of the time, produced a book entitled *Celestial Scenery*, in which he wrote: 'We have no valid reason to conclude that the degree of heat on the surface of different planets is inversely proportional to the squares of their distances from the Sun . . . There may be as much warmth experienced in that distant region of the Solar System as in the mildest parts of our temperate zones.' In other words, Uranus could be as hot as the Earth.

Obviously ideas of this kind could not survive for long, but very little could be found out about Uranus from sheer telescopic observation. The first scientific reports of markings on the disk – two round white spots – came from J. Buffham in January 1870. Buffham had a good reputation, and he was using a power of × 320 on a 9-inch refractor, but whether or not the spots on Uranus were genuine seems dubious. Vague belts, zones and shadings were reported by

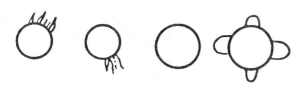

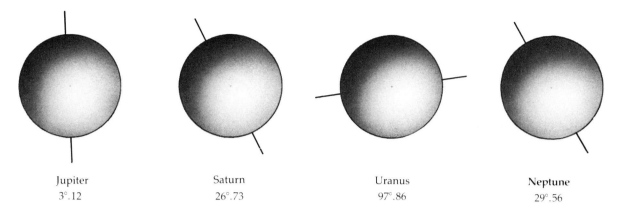

Jupiter	Saturn	Uranus	Neptune
3°.12	26°.73	97°.86	29°.56

later observers from time to time, but their reality was always suspect. In 1970, a century after Buffham's report, photographs of Uranus were taken from a telescope carried up to 24 kilometres in the balloon *Stratoscope II*. The resolution was given as 0.15 seconds of arc, ten times better than anything which could be achieved with a ground-based telescope, but the Uranian disk appeared to be completely blank.

Until our own century it was still thought possible that the giant planets could be capable of sending appreciable heat to their families of satellites, but all such ideas were disproved in the 1920s by a series of brilliant papers from Sir Harold Jeffreys. It had become clear that the visible surfaces are gaseous; it had also become evident that the outer clouds are bitterly cold. The mean surface temperature of the upper layers of clouds which cover Uranus is about − 210 degrees Centigrade. William Herschel's planet is indeed a chilly world.

Above *The axial inclination of the major planets (excluding Pluto).*

Below *Uranus photographed by a balloon-borne telescope from an altitude of 80 000 feet above the surface of the Earth. This photograph is a composite of 17 images obtained from the Stratoscope II system by Bob Danielson and Martin Tomasko in 1970.*

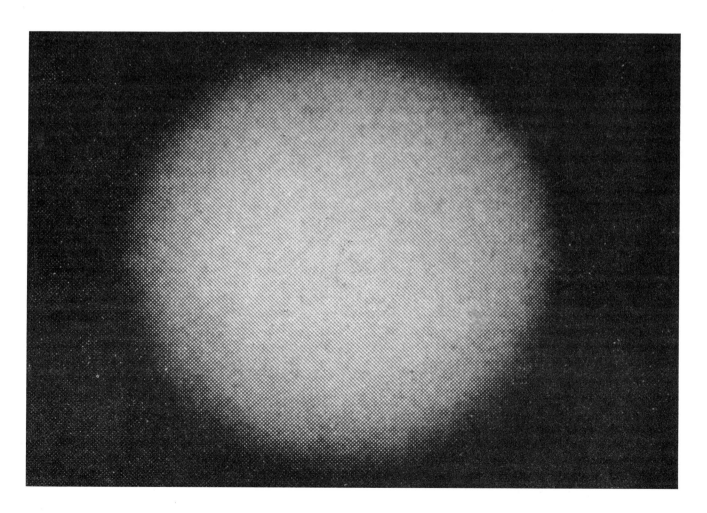

The rotation of Uranus

With a planet which shows visible surface markings, measuring the rotation period is fairly straightforward. This is the case with Mars, and Herschel used the dark features to give a period which was very close to the true value of 24 hours 37 minutes 22.6 seconds – though admittedly his observations were later re-analysed by the German astronomers Wilhelm Beer and Johann von Mädler; Herschel's own analysis had been less accurate.

Mercury and Venus are much less easy to study, and it is only in modern times that their rotation periods have been found: 58.6 days for Mercury, 243 days for Venus – with the added complication that Venus rotates in an east–west or retrograde direction.[†] Jupiter is different. At the level of the cloud-tops the rotation period is less than ten hours, but is not the same for the whole of the planet; there is a strong, fast-moving equatorial current, and various discrete features such as the Great Red Spot have periods of their own, so that they drift around in longitude. The Great Red Spot has been found to be a whirling storm – a phenomenon of Jovian 'meteorology' – and may eventually die out; it is less conspicuous now than it used to be a hundred years ago.

Saturn's surface features are much less evident than those of Jupiter, partly because Saturn is smaller and further away but mainly because it is genuinely less active. Again the spin is rapid, and is everywhere less than 11 hours. But Uranus presents problems of its own, and ordinary telescopic observation could be of little help.

The axial inclination of 98 degrees to the perpendicular (more than a right angle) was not questioned; technically the rotation is in the retrograde sense, and the same applies to the orbital motions of the satellites. There are times when one or other of the poles of the planet is presented to the Sun, and so appears in the middle of the disk as seen from Earth. This is true at the present moment; in 1985 the south pole was presented, while by 2030 it will be the turn of the north pole to have a 'midnight sun' lasting for many Earth-years. At other times, as in 1966, the equator is presented. To find out the rotation period under such conditions is obviously very difficult, particularly in view of the lack of well-marked cloudy features.

In 1856 the French astronomer J. Houzeau carried out a theoretical investigation, based upon the shape and possible internal constitution of Uranus. He came to the conclusion that the rotation period was likely to be between 7.25 and 12.5 hours, of the same order as those of Jupiter and Saturn, but it must be admitted that this was little more than an intelligent guess. Slight variations in brightness might be more informative, and various measurements were made over the years, but Uranus does not fluctuate much – by no more than a few hundredths of a magnitude over short periods – and again the results were unreliable. They ranged from 10 hours 50 minutes (E.C. Pickering) through to 21 days (R.L. Waterfield). It was all very uncertain, but better results might be expected from spectroscopic observations of the well-known Doppler effect. This research was initiated in 1902 by Henri Deslandres, in France.

Most people are familiar with the Doppler effect in one form or another. If a police car or ambulance speeds by, sounding its alarm signal, the note will be higher-pitched when the vehicle is approaching than when it is receding – because sound is a wave-motion, and with an approaching source more sound-waves enter the listener's ear than would otherwise be the case, so that the wavelength is effectively shortened; with a receding source the wavelength is effectively lengthened. The same principle applies to light. An approaching source will seem a little 'too blue', a

---

[†] The reason for Venus' slow, retrograde rotation is still not known with certainty; the period is longer than that of the period of revolution round the Sun, which amounts to 224.7 days. Further details are given in *The Planet Venus*, by the present authors (Faber and Faber, London 1982).

The changing position of the poles of Uranus.

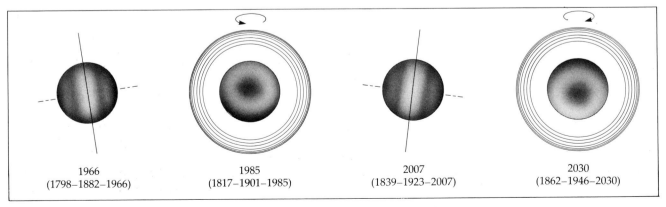

| 1966 | 1985 | 2007 | 2030 |
| (1798–1882–1966) | (1817–1901–1985) | (1839–1923–2007) | (1862–1946–2030) |

Drawing of Uranus made by R.L. Waterford following his observations of the planet with the 25-centimetre Cooke refractor at the Four Marks Observatory. The first two drawings relate to observations in 1915 and the latter to an observation in 1916. All the drawings show a series of faint belts reminiscent of Jupiter and Saturn.

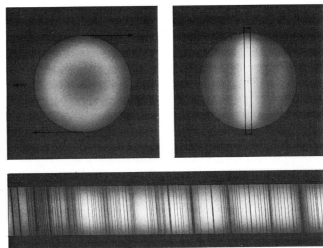

The rotational Doppler effect.

receding source 'too red' – since the wavelength of red light is longer than that of blue. Of course, the actual change in colour is inappreciable in everyday experience, because light travels so fast (3000 000 kilometres per second), but astronomically the Doppler effect is all-important.

Stars, including the Sun, are self-luminous. When their light is analysed by means of a spectroscope, normal stars show rainbow bands crossed by dark lines, each line being the particular trade-mark of one particular element or group of elements. Since a planet shines by reflected sunlight, it will show what is essentially a solar spectrum, but superimposed upon it will be dark lines due to the atmosphere of the planet itself. This was found to be true of Uranus. Dark spectral lines of Uranian origin were first seen as long ago as 1869 by the Italian pioneer Angelo Secchi.

Now consider a body which is rotating. One limb will be approaching us, assuming that the orientation is conventional, and will show a blue shift; the other limb will be receding, and the shift will be to the red. The effect can be used to measure the direction and rate of the rotation. Deslandres found that the rotation of Uranus is retrograde, as would be expected from the orbital movements of the satellites, and his work was fully confirmed in 1911 by Percival Lowell and V. M. Slipher at Flagstaff in Arizona, using the famous Lowell refractor. The period was given as 10 hours 49 minutes.

Admittedly this explanation is oversimplified, and there are many complications to be taken into account, but until the Voyager 2 mission the spectroscopic method was the best available. Later measurements indicated that Lowell's period was decidedly too short. In 1977 R. Brown and R. M. Goody made more detailed spectroscopic observations, and derived a period of 16 hours 19 minutes, with an uncertainty of 16 minutes either way; the *Astronomical Almanac* for 1981 gave 15 hours 36 minutes. It was left to Voyager 2 to give a final answer. The true period is 17 hours 48 minutes, with an uncertainty of less than one minute.

Obviously we do not have details about any differences in period between the equatorial regions and higher latitudes; Voyager 2 has gone on its way toward a rendezvous with Neptune in 1989, and we are once more reduced to ground-based observations. But at least we are sure that the mean value is correct.

This is of little help in accounting for the extraordinary inclination of the axis, which is so utterly unlike that of any other planet. Something very unusual must have happened early in Uranus' history. Assuming that the planets grew by accretion from the original solar nebula, and that this nebula was in a state of rotation, it would be logical to assume that all the planets would spin in the same sense – yet Venus does not, and Uranus is just is baffling.

Various explanations have been proposed, none of which is very convincing. One idea often favoured is that at a reasonably early stage in its evolution Uranus was struck by a comparatively massive body, and was literally tipped over. But when it is remembered that Uranus is over 50 000 kilometres in diameter, with a relatively massive core possibly overlaid by a superpressurized water/ammonia ocean beneath its extensive atmosphere, the concept of a collision of sufficient violence seems distinctly far-fetched. It would certainly have destroyed any existing satellites, in which case it is necessary to suppose that the present satellites were formed later.

There, for the moment, the matter rests. Despite the space-probes, the planets guard their secrets well.

Discovery of the rings

Once a planet has been found to have a satellite system, its mass can be found. More accurately, the movements of the satellites give the combined mass of the primary and the satellites combined – but in most cases, as with Uranus, the satellites are so lightweight that for most practical purposes their own masses can be neglected. The mass of Uranus is 14.6 times that of the Earth, and the escape velocity is 22.5 kilometres per second. Because of the overall lower density of the globe, the surface gravity would be found to be only 1.17 times that on the surface of the Earth – if anyone could get there and stand upon a layer of gas!

Measuring the diameter was much more of a problem, and it was important to know the exact size so that models of the internal structure could be worked out. It might be thought that the simple solution would be to look at Uranus through a telescope, use a micrometer or some equivalent device, and see how large the disk appeared, in which case its real diameter would follow. Unfortunately this is not so easy as it may sound. An object such as Uranus, with an apparent diameter of below 4 seconds of arc, will have a somewhat 'fuzzy' edge, and the slightest uncertainty will involve a large potential error. For example: in 1970 astronomers as Princeton University gave a value of 51 800 kilometres for the diameter of Uranus, while in 1977 J. Janesick at Mount Lemmon Observatory, also in the United States, gave 55 000 kilometres.

Another possible method was devised by Gordon Taylor, of the Royal Greenwich Observatory at Herstmonceux. He proposed to take advantage of the fact that even a body with a small apparent diameter, moving across the sky, will occasionally pass in front of a star, and hide or occult it. Lunar occultations are common; those by planets much less so – and the smaller a planet looks, the less the chance of its producing an occultation. However, Taylor found that the method worked well for some asteroids, and also for Jupiter's satellite Io. All that had to be done was to measure the duration of the occultation, observing from several different sites on the Earth's surface, and then use the results to find the diameter of the occulting body. The most irritating factor was that one had to wait for Nature to provide an occultation.

As long ago as 1952 Taylor started to search for future occultations by Uranus, using planetary tables together with the best star catalogues. The first real chance seemed to be due on 10 March 1977. This seemed to mean a long wait, and moreover, Uranus was moving in a rather barren part of the Zodiac, so that faint stars were inconveniently few.

The selected star was known merely by its catalogued number of SAO 158687. Its magnitude was 8.8, so that it was bright enough to be seen with a small telescope or even good binoculars; it was only about three magnitudes fainter than Uranus. It was an orange star, though in fact its colour made no difference to the occultation technique.

Taylor's calculations indicated that the occultation was likely to be seen from a track across the Earth's surface well south of the equator, so that observatories in countries such as Australia and South Africa would have the best opportunities. But the slightest error in either the predicted position of Uranus or the actual position of the star would make a great deal of difference; it needed only an error of half a second of arc to shift the limit of the occultation track from the equator to the pole.

On 22 January 1977 Uranus passed about 0.8 minutes of arc south of the star, and careful measurements were made in readiness for the March occultation. By then several major expeditions had been planned to observe it; for example French astronomers were going to South Africa, while American expeditions were scheduled for parts of Australia. There was also the Kuiper Airborne Observatory or KAO, named in honour of one of the pioneers of space exploration by unmanned probe – the late Gerard Kuiper, founder of the Lunar and Planetary Laboratory in Arizona. The KAO is essentially a C-141 aircraft which can fly at up to

The occultation of SAO 158687 by Uranus and the rings on 10 March, 1977. This observation led to the discovery of the rings of Uranus.

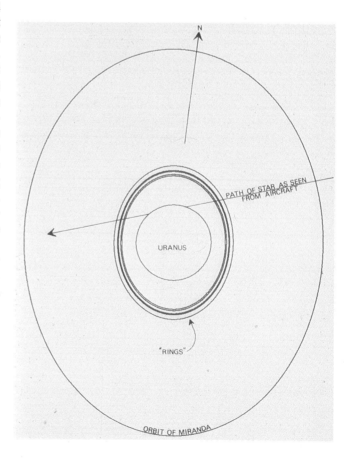

The Gerard P. Kuiper Airborne Observatory (KAO) carries a 0.9 metre infrared telescope to altitudes of 12 kilometres. From this height, above 85% of the atmosphere and more than 99% of the water vapour, astronomers are able to make infrared observations that are impossible at ground level.

12 kilometres above ground level, and which carries a 90-centimetre reflecting telescope peering out of the port side. The aircraft is extremely steady, and the ability to move around at will is an enormous advantage in observations such as that of the occultation by Uranus.

Consternation ensued when the January measurements showed that there had indeed been an error. It appeared that the track might turn out to be much further south than had been expected. Fortunately the programme went ahead – and the results were far more dramatic than anybody had dared to hope.

At 10.37 p.m. on 10 March the KAO took off from Perth International Airport. On board were three astronomers: J. Elliot, E. Dunham and D. Mink. The telescope was equipped with a sensitive photometer; it was expected that the chart pen would trace a constant level until the occultation began, when the signal strength would drop to that of Uranus on its own instead of that of Uranus and the star combined. When the star reappeared from behind the planet, the signal

would return to its former value, and the shape of the light-curve would yield information about the exact size and shape of Uranus itself. Ground-based observations were being carried out at the Cape in South Africa, in Perth, and at two stations in India and one in China, as well as at the Lowell Observatory in Arizona.

On board the KAO, observations began well before the predicted time of occultation. Quite unexpectedly the star 'winked' five times, 35 minutes before the occultation was due, and it seemed to the observers that these 'winks' might be due to some material in the neighbourhood of Uranus itself. Hastily they alerted the ground stations at Perth and Cape Town. The occultation took place on schedule, at 20.52 hours GMT, and lasted for 25 minutes; after emersion there were more 'winks', and it was found that the second series was symmetrical with the first. There could be only one possible explanation. Uranus must have a system of rings.

This was a complete surprise. Saturn's ring-system had been regarded as unique (the thin, dark ring of Jupiter was not found until the rendezvous with Voyager 1, in March 1979) and nothing of the sort had been anticipated for Uranus. But there could be no doubt; confirmation, again by the occultation method,

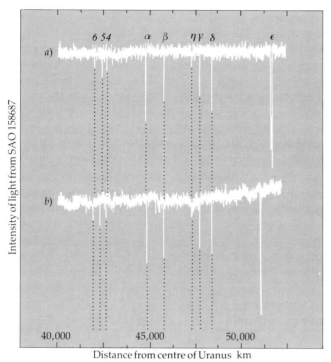

The drops in intensity of light from the star SAO 158687 as it passed behind the rings of Uranus on March 10 1977.

The first clear image of the rings of Uranus. Computer processing creates a false three-dimensional appearance, but it does allow the dark ring system to be seen near the much brighter planet. The collective ring system is shown, as the individual rings could not be resolved.

The rings of Uranus.

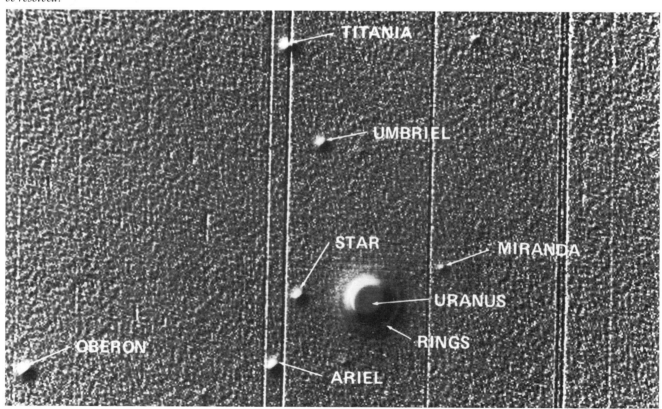

was soon forthcoming (Uranus, for once, was obliging in its wanderings), and in 1978 astronomers at Palomar, using the 200-inch Hale reflector, managed to obtain a photograph. Subsequently a much better image was obtained in infra-red light by D. A. Allen and his colleagues at the Anglo-Australian Observatory in New South Wales. Well before the fly-by of Voyager 2, in January 1986, the existence of nine, dark, narrow rings had been established. They were studied by two different teams, each of which had its own nomenclature – one using numbers from 'inside to outside', the other using Greek letters. The resulting compromise on nomenclature meant that reckoning outward from Uranus, the rings became known as 6, 5, 4, alpha (α), beta (β), eta (η), gamma (γ), delta (δ) and epsilon (ϵ). Following Voyager 2, and a much more thorough knowledge of the rings, this chaotic system will no doubt be changed before long.

Predictably, the rings lie in the plane of the Uranian equator. Less predictably, not all of them are circular, and the outermost, the ϵ ring, is also irregular in width. While Saturn's rings are as bright as snow, those of Uranus prove to be as black as coal-dust, so that their nature is completely different.

It may be said that the discovery of the ring-system was fortuitous, but this would be unfair both to Gordon Taylor and to the teams of observers monitoring the occultation of SAO 158687. Of course the rings would have been found by Voyager 2 in any case; the fact that they were detected earlier is a tribute to the skill and ingenuity of those involved.

Has Neptune, too, a ring? The occultation method has been tried; the observations are even more difficult than with Uranus, because the apparent diameter of Neptune is never more than 2.2 seconds of arc. Moreover, Neptune is unique in having a large satellite (Triton) with retrograde motion, whereas the axial rotation of Neptune itself is conventionally direct. Conditions close to the planet may therefore be unsuited to the existence of a ring-system. We cannot be sure, and the results so far have been inconclusive; the possibility of incomplete rings or 'arcs' has been suggested, but there is nothing really positive. Once again we have to pin our hopes on Voyager 2. If it passes Neptune in August 1989 and functions as efficiently as it did during the rendezvous with Jupiter, Saturn and Uranus, we will find out whether or not Neptune is the only ringless giant. If Voyager 2 fails, then we may have to wait for many years before we know.

Uranus and its rings, photographed in infra-red by D.A. Allen with the 3.9-metre Anglo-Australian telescope at Siding Spring in 1984. Because methane gas dims Uranus in the longer red wavelengths, the planet appears blue-green, while the rings that encircle it show up as dull red. This was the first really good Earth-based picture of the rings ever taken.

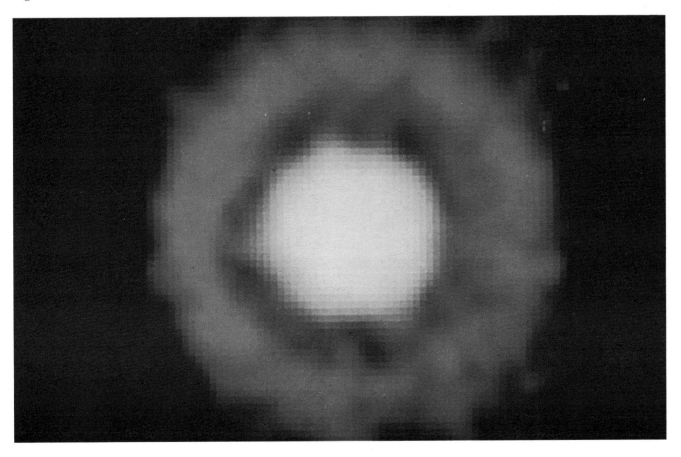

Space-probes to the planets

The idea of travelling to the Moon and planets is very old. The first science-fiction novel about a lunar voyage dates from about AD 150, and was the work of a Greek satirist, Lucian of Samosata; he did not mean to be taken seriously (he called his book the *True History* because, in his own words, it was made up of nothing but lies from beginning to end!) but later writers thought differently. Johannes Kepler's *Somnium* or Dream, published posthumously in 1634, gave a description of the Moon (and its inhabitants) as he believed the lunar world to be, even though his method of transport – demon power – would hardly have recommended itself to NASA. Perhaps the most influential of all Moon-voyage novels came in 1865 from the pen of the great French story-teller Jules Verne. It was called *From the Earth to the Moon* (the sequel, *Round the Moon*, followed a few years later) and in it Verne used scientific facts as accurately as he knew how. His main error was in using a space-gun instead of a rocket to take his travellers from one world to another.

It was not until near the end of the last century that the first practical proposals were made. They were due to be a shy, deaf Russian school-teacher named Konstantin Eduardovich Tsiolkovskii[†], who realized that the only way to travel in space is to use the principle of reaction: 'every action has an equal and opposite reaction'. This involves rocketry, and only rockets can function in space, where there is no air or other surrounding medium. He was purely a theorist, and never carried out any experiments himself, but many of his ideas were sound, and in particular he advocated the use of liquid propellants rather than solid fuels. His work caused little interest when first published, possibly because it came out in obscure Russian journals, but by the time he died, in 1934, he had received considerable recognition for his pioneering efforts. In the USSR today he is rightly regarded as 'the father of space-travel'.

In 1919 Robert Hutchings Goddard, in the United States, published a paper entitled *A Method of Reaching Extreme Altitudes*, in which he discussed the possibility of sending a small vehicle to the Moon. He was ridiculed by the Press, but in 1926 he successfully fired the first liquid-propellant rocket in history. It was modest in both size and performance, but it worked, and interest began to grow.

The first interplanetary society was formed in Germany in 1927, and rockets were developed, at first for scientific purposes and later – when the Nazis took over – for military weapons. At Peenemünde, in the Baltic, the group masterminded by Wernher von Baun

produced the V2 weapons used to bombard London during the final stages of the war; the campaign ended only when the launching sites were overrun in 1944. The V2 was liquid-propelled, and must be classed as the direct ancestor of Voyager 2's launcher rocket. After the German collapse, von Braun went to the United States, and it was he who was mainly responsible for sending up the first American satellite, Explorer 1, in 1958. But by then the Russians had taken the lead; and within two years from the launching of Sputnik 1 in October 1957, unmanned Soviet vehicles had reached the Moon. As we all know, the Americans quickly recovered their lost ground, and in July 1969 Neil Armstrong stepped out on to the surface of the Moon.

The Moon is our faithful companion; the planets, moving round the Sun rather than round the Earth, are much more difficult targets, quite apart from their being much further away. Venus was first on the list, with Mariner 2 in 1962; then came Mars, and then Mercury. The Mariner 10 flight to Mercury is of special interest in view of the Voyager missions. Mariner 10 was launched from Cape Canaveral on 3 November 1973, and flew past Venus on 5 February 1974, using the gravitational pull of that peculiar planet to send it on to a rendezvous with Mercury on the following 29 March. The probe was then put into an orbit which took it back to the neighbourhood of Mercury in September 1974 and again in 1975, but after the third active encounter its transmitters failed and all trace of it was lost – though no doubt it is still in solar orbit, and still making regular approaches to Mercury.

This was the first time that the technique of 'gravitational assistance' had been used – a procedure which has been rather irreverently nicknamed interplanetary snooker! But when missions to the outer giants were considered, the technique became more or less essential unless the time-scale were to be inconveniently long. As we have already noted, Nature was helpful inasmuch as during the late 1970s the four giants were suitably 'lined up' in a curve, and this led to the concept of the Grand Tour, with probes passing Jupiter, Saturn, Uranus and Neptune in turn. The plan had to be modified several times, mainly because of financial cutbacks in NASA's budget – and no other nation (not even Russia) was at that time capable of anything so ambitious. It is probably true to say that Voyager 2 achieved the Grand Tour in the end almost without the politicians realizing it.

At the moment (1986) the American planetary programme is suspended. Various probes had been planned for the late 1980s, including a Venus radar mapper, a new mission to Mars, and the 'Galileo' mission to Jupiter. Galileo was perhaps the most important. It was to consist of an entry probe, which would plunge into Jupiter's clouds and send back information for as long as possible before being

---

[†] The name may be spelled in various ways: 'Tsiolkovsky' is another, while the historian of science, Willy Ley, preferred the phonetic and much more sensible 'Ziolkovsky'.

destroyed, and an orbiter, which was to be put into a closed path round the planet and make long-term studies of the surface, plus regular close approaches to the satellites; on the way to Jupiter, Galileo was also scheduled to make the first rendezvous with an asteroid (Amphitrite). Sadly, all these were geared to the Shuttle, and the disaster with *Challenger* has led to an indefinite delay. We cannot now hope for any further launchings to the planets until well into the 1990s, which makes Voyager 2 more important than ever.

Other nations have entered the field of space research. China is making rapid progress; Japan sent two very successful small probes to Halley's Comet in March 1986, and, of course, there was the triumphant Giotto, from the European Space Agency, which was launched from Kourou, in French Guyana, but was built mainly in Bristol by British Aerospace. Giotto survived its passage through the heart of Halley's Comet, and is still under control, through inevitably it has been damaged. The Soviet Vega probes to Halley's Comet were also highly successful, and we cannot rule out the chance of a major Russian mission to the outer planets well before the American programme resumes. But this lies in the indefinite future, so let us turn to the Voyagers, which have already achieved so much and whose work is by no means over yet.

Voyager 2 was launched, from the John F. Kennedy Space Center on August 20, 1977, atop a Titan-Centaur booster rocket.

The Voyager space-craft

Voyagers 1 and 2 are to all intents and purposes identical space-craft. Voyager 1 was launched from Cape Canaveral on 5 September 1977; it by-passed Jupiter on 5 March 1979 and used the powerful Jovian gravity to send it on to Saturn on 12 November 1980. It is now on a path which has taken it well out of the plane of the ecliptic, so that it will have no more planetary encounters (unless we count the hypothetical Planet X, which may or may not exist, and about which more will be said later). Engineers are still in regular touch with it, and it continues to provide unique information about regions in which no space-craft has ever travelled before. We may even maintain contact with it until it reaches the 'heliopause', the region where the solar wind ceases to be detectable.

Voyager 2 was actually launched first, on 20 August 1977, but was moving in a less economical path; it passed Jupiter on 9 July 1979 and Saturn on 25 August 1981. The success of Voyager 1 in making a close-range study of Saturn's satellite Titan enabled the space-craft controllers to alter the trajectory of Voyager 2 so that its path would lead to an encounter with Uranus in January 1986 and, ultimately, Neptune in August 1989. The original objectives of the mission had therefore been greatly expanded, to include the first fly-bys of the most distant known planets in the Solar System.

The Voyager space-craft are based upon the highly successful Mariner design used on so many earlier reconnaissance missions to the inner planets. Each Voyager weighs 825 kilograms (1819 pounds), carries eleven scientific instruments, and is dominated by the 3.66-metre (12-foot) antenna and the huge 10-metre magnetometer boom. The on-board computer systems are capable of directing the scientific instruments and the engineering equipment which controls the space-craft operation. The Voyagers carry their own source of electricity in the form of three Radioisotope Thermo-electric Generators (RTGs), since there is insufficient sunlight in the depths of the Solar System to provide any additional energy. The RTGs are miniature nuclear power-plants which convert about 7000 watts of heat into some 400 watts of electricity. At launch, the power output from the RTGs was 475 watts, but this decreases at a rate of about 7 watts per year due to the half-life of the fissionable plutonium oxide and the degradation of the silicon-germanium thermocouples. However, there is more than enough power for the Voyagers to function beyond the year 2000.

The Voyager 2 space-craft. This picture shows the full size engineering model located in the Von Kármán Auditorium at the Jet Propulsion Laboratory, Pasadena, California.

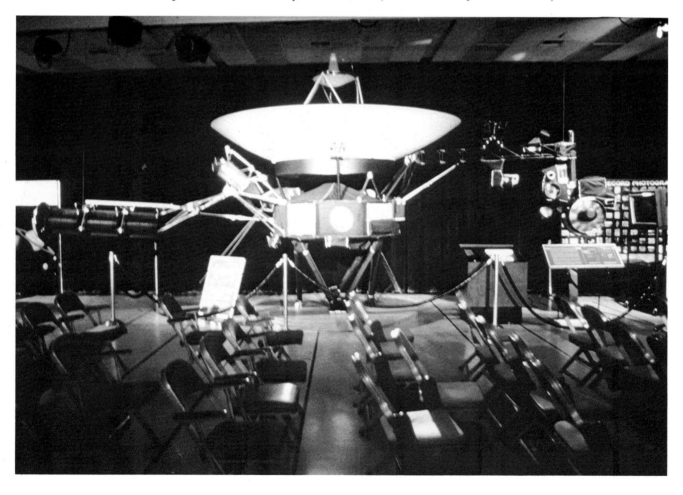

In flight, each space-craft is three-axis stabilized, using the Sun and a bright star as the celestial references. Behind the antenna is the main 'bus', a ten-sided aluminium framework containing the space-craft's electronics. This structure surrounds a spherical tank containing the hydrazine which is used for manœuvring the space-craft in flight. Although the Voyager is a very complex machine, it must be told what to do and what sensor to use, and a set of instructions generated in advance to tell the sensor what to do and when to do it. We must remember that the space-craft is rapidly receding from the Earth, and at Uranus the one-way communication time from Mission Control at the Jet Propulsion Laboratory in Pasadena to Voyager 2 is about 2 hours 45 minutes. The three key sub-systems for controlling the engineering aspects of the mission and supervising the scientific investigations are the Computer Command Sub-system (CCS), the Flight Data Sub-system (FDS) and the Attitude and Articulation Control Sub-system (AACS).

The CCS is the brain of the space-craft, and has two main functions: to carry out instructions from the ground to operate the space-craft, to perform the housekeeping functions and to gather the science data. It must also be alert for any problem with, or malfunction of, the numberous sub-systems and to be able to respond immediately. It consists of two computers, each having a 4096-word (18 bit words) memory. Although both are currently used in a non-redundant

fashion, the mission could still operate even if one of them failed. The observational plan for the entire encounter period requires about 18 000 CCS words. For example, about 275 CCS words are required to perform a trajectory correction manœuvre. However, these available words have to be divided between the various time periods of the encounter, which may extend over several months.

There is a two-way communication system with the Voyager space-craft; the uplink contains the command data, while the downlink contains the science and engineering telemetry data. The engineering data are transmitted at 40 bits per second (bps) at S-band, and are also embedded in the X-band science data transmitted at the higher data rates, varying between 4.8 and 21.6 kilobits per second. The FDS, which consists of two reprogrammable digital computers, is responsible for the collection and formatting of the telemetry data. For the Uranus encounter, it is also responsible for some data processing of the imaging data, through the operation of the data compression algorithms which are necessary to reduce the volume of data finally transmitted to the ground stations. Before the Uranus encounter, one of the FDS computer memories lost a block of 256 memory words out of a

The flight paths of the Voyager space-craft through the outer solar system.

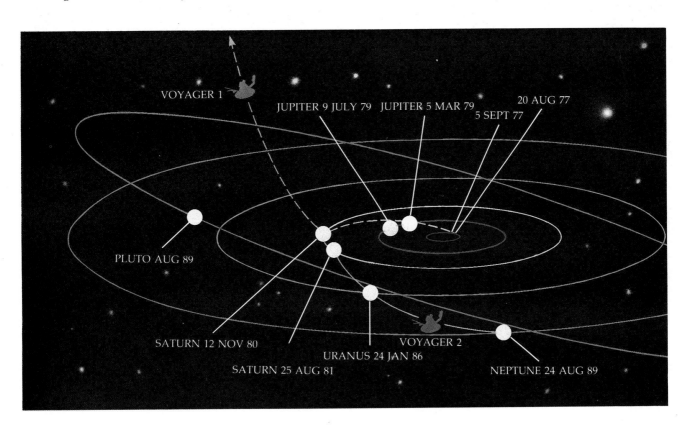

total of 8192. This must be regarded as a relatively minor failure. However, the loss of a further 512 words from the primary memory would be more serious, and would then result in the need to abandon the dual processor mode of the system.

There are two classes of space-craft, spin-stabilized and three-axis-stabilized. The former configuration, which has been used for the Pioneer 10, 11 and Venus missions, stabilizes the space-craft by spinning, so that the entire probe acts as a gyroscope. The three-axis-stabilized space-craft, such as the Voyagers, maintain a fixed orientation in space except when manœuvring. This stabilization is maintained by the AACS computer system, which also controls the motion of the scan platform. Gyros may be used for special purposes and short periods of time (a few hours) to maintain the attitude of the probe. In the celestial control mode, the Voyagers maintain their fixed attitude in space by viewing the Sun and a bright star such as Canopus, Fomalhaut or Achernar. If the space-craft should drift from its proper orientation by more than an agreed amount, then the AACS will issue commands to fire the hydrazine jets to correct the attitude.

There are occasions when the Voyager cannot send science data direct to the Earth – for example, during a manœuvre, or when the probe is behind a planet. In these situations a digital tape (DTR) is used to store the data, and play back the results to Earth at a more convenient time in the mission. The DTR has eight tracks, each of which can hold an amont of data equivalent to 12 images (6MB). Normally, the DTR is shared between all the science experiments; it is operated at three speeds – 115.6 kilobits per second (record), 21.6 kilobits per second (playback) and 7.2 kilobits per second (record and playback).

The eleven scientific instruments consist of ten experiments and the space-craft radio system. They can be divided into two general classes: particle and field instruments, which are body fixed, and directional instruments concerned with remote sensing of the atmospheres, satellites and ring-systems of the target planets. There are five directional sensors: imaging science with the narrow and wide angle television cameras (ISS), the infra-red interferometer spectrometer radiometer (IRIS), the photopolarimeter (PPS) and the ultra-violet spectrometer (UVS). All of these instruments are mounted on the steerable scan platform. The remaining six instruments measure the energetic particles, radio emissions, and the magnetic fields in both interplanetary space and in the individual planetary systems. This set of instruments represents the most complete set of experiments yet flown in the exploration of the Solar System. So far, it has worked amazingly well.

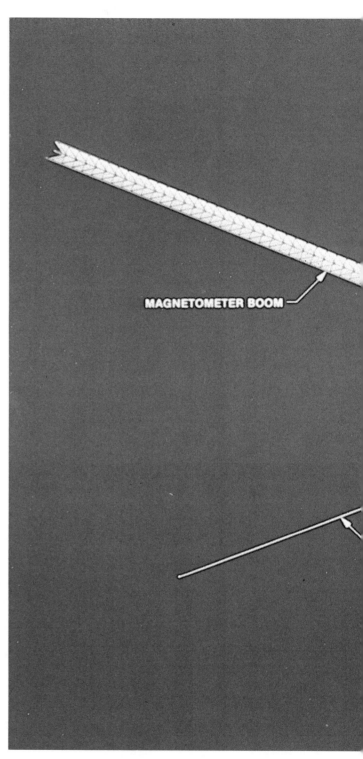

The Voyager 2 space-craft and its instruments.

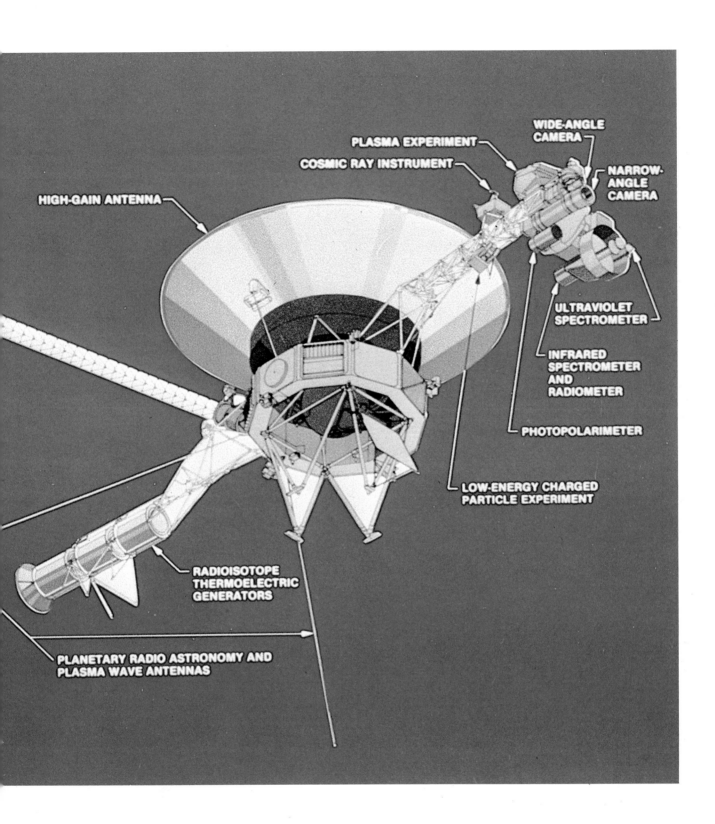

PLASMA EXPERIMENT

COSMIC RAY INSTRUMENT

WIDE-ANGLE CAMERA

NARROW-ANGLE CAMERA

HIGH-GAIN ANTENNA

ULTRAVIOLET SPECTROMETER

INFRARED SPECTROMETER AND RADIOMETER

PHOTOPOLARIMETER

LOW-ENERGY CHARGED PARTICLE EXPERIMENT

RADIOISOTOPE THERMOELECTRIC GENERATORS

PLANETARY RADIO ASTRONOMY AND PLASMA WAVE ANTENNAS

Earlier results from the Voyagers

The Voyagers were not the first probes to the outer part of the Solar System. They had been preceded by Pioneer 10, which surveyed Jupiter in December 1973, and Pioneer 11, which made its pass of Jupiter in December 1974 and was then diverted back across the Solar System to a brief and initially unscheduled encounter with Saturn on 1 September 1978. However, in view of the unique difficulties of the Voyager mission to Uranus, it may be helpful to say a little about what had already been discovered about the two closer giant planets.[†]

Radio emissions from Jupiter had been detected as early as 1955 (admittedly by accident), and it was believed that there must be a powerful magnetic field, together with radiation belts of a type similar to the Van Allen zones surrounding the Earth. Pioneer 10 passed by Jupiter at a distance of 132 000 kilometres, and showed that these speculations had been correct; in fact the radiation belts were 10 000 times more intense than ours, and the Pioneer instrumets were 95 per cent saturated. If the approach distance had been much less, the equipment would have been put out of action altogether. Pioneer 11, already on its way, was given a course correction to take it quickly over Jupiter's equatorial zone, where the radiation levels were highest. The Jovian magnetosphere was found to be so extensive that the 'tail' extends out beyond the orbit of Saturn, so that at times Saturn may lie inside it.

[†] For full details see *The Solar System*, edited by the present authors (Mitchell Beazley, London 1982).

The magnetosphere of Jupiter.

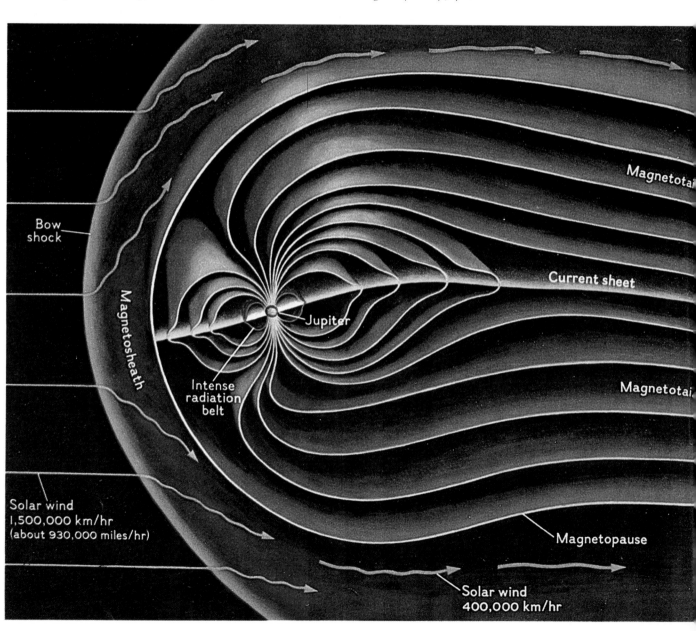

Above *The satellites of Saturn shown to scale;* from the top downward, reading from left to right. *Mimas, Enceladus, Tethys, Dione, Rhea, Iapetus, Hyperion and Phoebe.*

sometimes outside it. Radiation zones exist with Saturn, but are less intense than those of Jupiter. The numbers of electrons fall off sharply at the outer edge of the main ring-system (that is to say, the outer edge of Ring A) because the electrons are absorbed by the icy ring particles.

Auroræ on the night sides of both Jupiter and Saturn were recorded by the Voyagers. So far as the disk features were concerned, superb images of the belts, zones, spots and other features were obtained, and wind velocities were measured. With Jupiter, the wind zones followed the boundaries between the bright and dark regions, as had been expected, but this was not true of Saturn; between latitudes 35 degrees north and 35 degrees south there is an eastward jet-stream with winds up to 1500 kilometres per hour, faster than any winds found on Jupiter.

Seasonal differences were found between the two hemispheres of Jupiter and Saturn. They were, naturally, more pronounced on Saturn, which has a much longer 'year'. The Sun crossed into Saturn's northern hemisphere in 1980, after the edgewise presentation of the rings to the Sun and to the Earth, but there is a pronounced lag effect, and during the Voyager missions the north pole was still about 10 degrees colder than the south pole.

The satellite systems were of immense interest. Jupiter has four very large satellites, known collectively as the Galileans (Io, Europa, Ganymede and Callisto), which are visible with a very small telescope; there is evidence that keen-sighted people can glimpse Ganymede at least with the naked eye, and a report of it goes back to Chinese times. In 1892 E. E. Barnard discovered a much smaller and closer-in satellite, Amalthea. Between 1904 and 1938 eight more small satellites were found, all well beyond the orbit of Callisto, outermost of the Galileans; these were divided into two definite groups, one (Leda, Himalia, Lysithea and Elara) at around 11 000 000 kilometres from Jupiter, and the other (Ananke, Carme, Pasiphaë and Sinope) at around 23 000 000 kilometres. The four members of the outer group have retrograde motion, and it is widely believed that all the satellites beyond Callisto are captured asteroids. Finally, Jupiter is now known to have three more small inner satellites; Metis and Adrastea (closer-in than Amalthea) and Thebe (between the orbits of Amalthea and Io).

Before January 1986 (or, to be more accurate, the end of December 1985, when Voyager 2 came within range of Uranus), five Uranian satellites were known, of which four were between 1000 and 2000 kilometres in diameter and the fifth below 500 kilometres. There was thus very little in common between the systems of Jupiter and Uranus. The Galileans are of planetary size; Ganymede is actually larger than Mercury and Callisto not much smaller, while Io and Europa are

Both the Pioneers obtained excellent images of Jupiter, and Pioneer 11 sent back some useful data about Saturn, but neither could compare with the much more sophisticated Voyagers. With the present if temporary collapse of the American planetary programme, it is fair to say that most of our detailed knowledge of the outer Solar System comes from the Voyager missions.

Saturn's magnetic field turned out to be a thousand times stronger than that of the Earth, though twenty times weaker than that of Jupiter. The magnetic axis was within one degree of the axis of rotation – a much closer alignment than with Jupiter – and, as with Jupiter, the magnetic poles are in the opposite sense to the geographical ones; thus the north pole of rotation has south polarity, and vice versa. Saturn's magnetosphere is variable in extent. Its average boundary lies close to the orbit of its principal satellite, Titan, so that Titan is sometimes inside the magnetosphere and comparable with our Moon. Yet they make up a varied family. Callisto and Ganymede are icy and cratered; Europa is icy and smooth, while Io has a violently active surface, with sulphur volcanoes in constant eruption – in appearance it has been likened to a red pizza! Amalthea is also reddish, and is irregular in shape, with a longest diameter of 270 kilometres. Himalia is 185 kilometres in diameter, all the rest are well below 100 kilometres.

The Saturnian system is different again. Here we have one very large satellite, Titan, and three pairs of icy satellites – Rhea/Iapetus (around 1500 kilometres in diameter), Dione/Tethys (around 1000 kilometres) and Enceladus/Mimas (400–500 kilometres). Also known in pre-Space Age times were Hyperion (irregular in form, with a longest diameter of 400 kilometres) and Phœbe (160 kilometres), which has a mean distance from Saturn of almost 12 000 000 kilometres, moves in a retrograde direction, and is almost certainly a captured asteroid.

New satellites, found within the last ten years, are very small. Atlas, Prometheus, Pandora, Epimetheus and Janus are closer in than Mimas; the last two move in virtually the same orbit, periodically exchanging paths, and are probably fragments of a former body which broke in two. Atlas acts as a 'shepherd satellite' to the A ring, keeping the ring particles in place, while Prometheus and Pandora do the same for the much fainter F ring. Two midget worlds, Calypso and Telesto, share an orbit with Tethys; Helene moves in the same orbit as Dione. There are no dangers of collisions, because the smaller bodies keep either 60 degrees ahead or 60 degrees behind the larger ones. (This principle is nothing new. The Trojan asteroids, of which many are now known, share the orbit of mighty Jupiter.)

Of the pre-Voyager satellites of Uranus, all were thought to be of approximately the same size as Dione or Tethys apart from Miranda, which was smaller, more comparable with Enceladus or Mimas. Therefore, it was believed that the Voyager findings at Saturn would give strong indications of what might be expected at Uranus. It was found that all Saturn's medium-sized icy satellites were heavily cratered apart from Enceladus, which turned out to have a very bright surface (the albedo or reflecting power is almost 100 per cent) with relatively small, fresh-looking craters, and wide areas which were crater-free altogether. On the other hand Mimas was dominated by one huge crater, appropriately given the name 'Herschel', which is 130 kilometres across – about a third the diameter of Mimas itself. If this crater had been formed by an impacting object, Mimas would probably have been broken up. Tethys also showed a crater, christened Odysseus, which is of similar relative size to Herschel, bearing in mind that Tethys is

considerably larger than Mimas; Tethys also showed strange trenches and troughs, one of which extended from near the north pole across the equator and into the far south – a total distance of 2000 kilometres. It was at least 4 kilometres deep, with a rim rising to half

Above right This computer enhanced image of Saturn shows the rings and their shadows of the illuminated crescent of the planet. Voyager 1 returned this image on November 13, 1980 at a distance of some 1,500,000 kilometres from the planet. The limb of the planet has been overexposed in order to bring out more detail in the rings. The inner region of the rings (the C-ring) scatters light in a way that causes it to look bluer than the outer rings (the B- and A-rings). The scattering allows a determination of the nature of the ring particles.

Below right The underside of Saturn's rings is shown in this Voyager 1 image. It was recorded on November 12, 1980, about eight hours after the spacecraft crosses from the northern to the southern side of the ring plane. In this image the rings are backlit and they look dramatically different to the images received earlier when the illumination was direct. The normally dark Cassini Division now shows as the brightest feature in the ring system.

The giant Saturnian satellite Titan, seen by Voyager 2 on the 23 August, 1981.

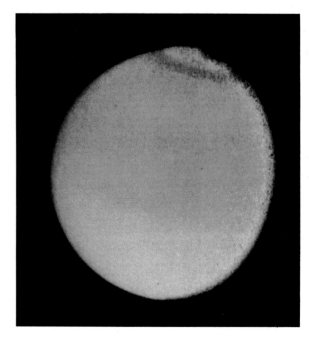

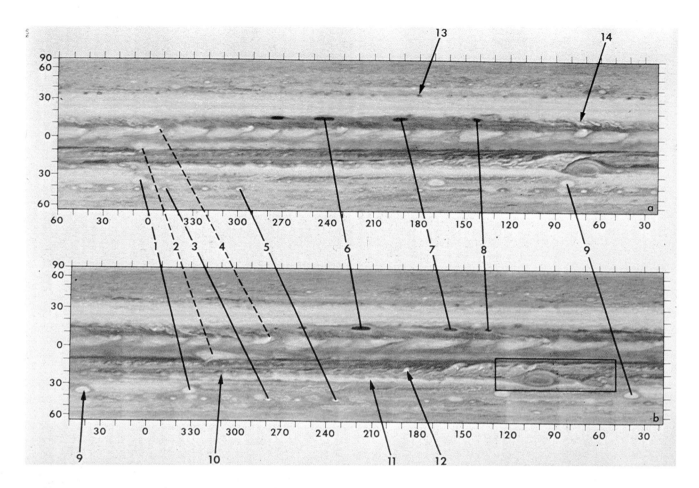

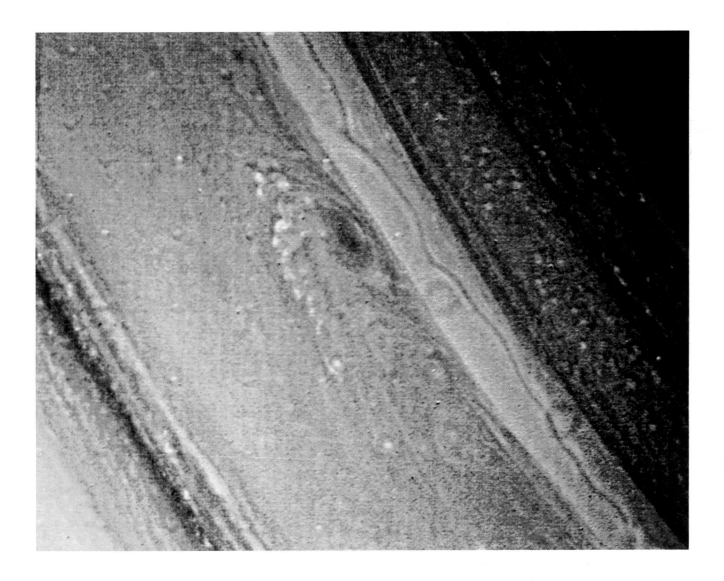

Above *Some cloud features seen in the atmosphere of Saturn during the Voyager fly-bys; above northern hemisphere, Voyager 2, 19 August, 1981, showing several cloud spots; overleaf northern hemisphere, Voyager 2, 20 August, 1981, showing a jet stream, ribbon-like feature and the generation of a cloud vortex.*

Top left *Cylindrical projections of Jupiter; top, Voyager 1 for the 1 February 1979 and bottom, Voyager 2 for the 23 May, 1979. From the comparison of the two images it is possible to follow the motions of the individual cloud features in this three month period.*

Bottom left *Saturn, as seen by Voyager 2 on the 4 August, 1981.*

Right *Voyager 1 image of Jupiter showing the familiar coloured bands of alternating light and dark cloud systems and the localised weather systems, such as the Great Red Spot and the systems of white cloud vortices.*

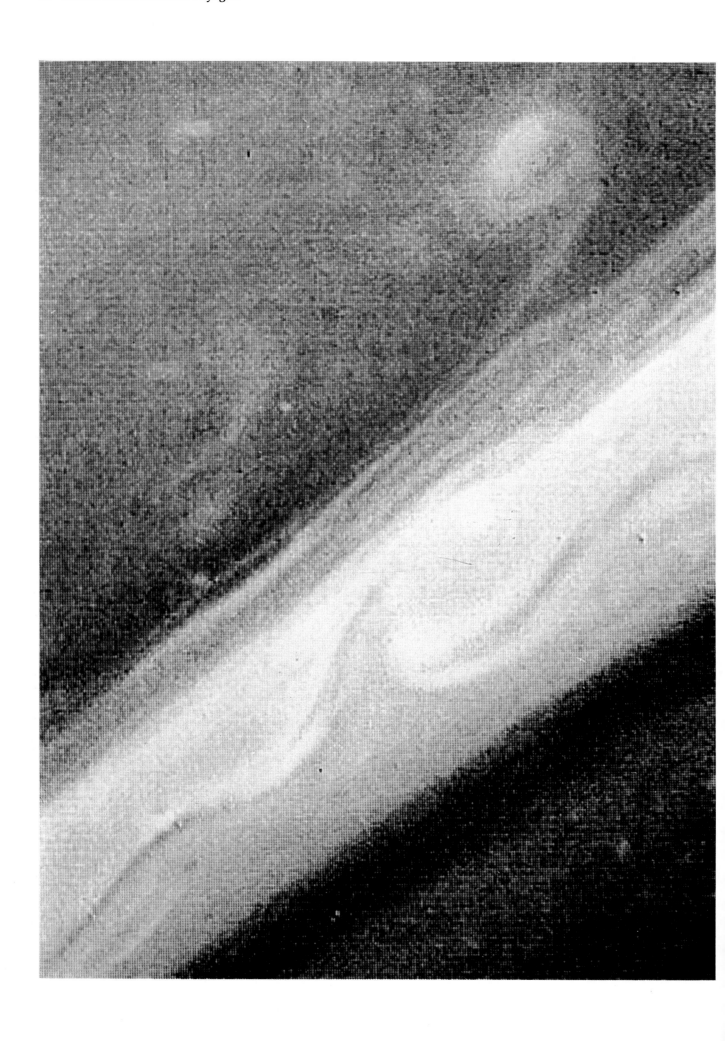

a kilometre above the outer surface. Dione, more massive than Tethys and slightly larger, had a darkish trailing hemisphere with a brighter leading hemisphere. (All the satellites, except the asteroidal Phœbe, have captured or synchronous rotations; that is to say, their axial rotation periods are the same as their orbital periods, as in the case of our own Moon.) The most prominent feature on Dione, now called Amata, is of uncertain nature; it may be either a crater or a basin, and is associated with a system of bright wispy features which may be due to ice which has seeped out from the interior of the satellite. Rhea is very heavily cratered, with a surface which gives every impression

of being ancient; Hyperion is darker, with one long ridge or scarp; Iapetus is puzzling inasmuch as it has one very bright hemisphere and one hemisphere which is extremely dark. This had already been inferred from telescopic observation, because Iapetus is much brighter when to the west of Saturn then when to the east, but in addition Voyager 2 showed that much of the leading hemisphere is covered with a dark

The family portrait of the Galilean satellites of Jupiter, Io, Europa, Ganymede and Callisto observed by the Voyager space-craft in 1979.

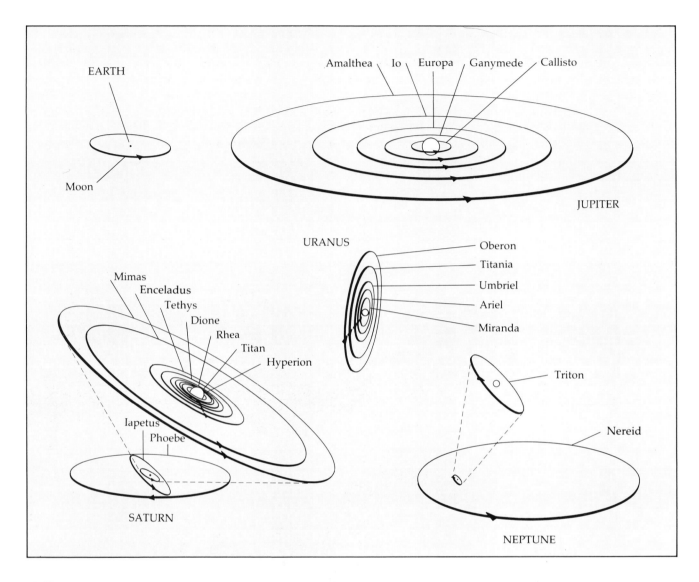

EARTH

Moon

Amalthea   Io   Europa   Ganymede   Callisto

JUPITER

URANUS

Oberon
Titania
Umbriel
Ariel
Miranda

Mimas
Enceladus
Tethys
Dione
Rhea
Titan
Hyperion

Iapetus
Phoebe

SATURN

Triton

Nereid

NEPTUNE

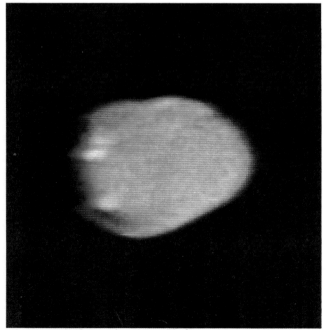

Above *The orbits of the major satellites of the solar system to scale.*

Left *The irregular shaped satellite Amalthea, first seen in detail during the Voyager 1 fly-by of Jupiter on the 4 March, 1979.*

Right *Comparison of the sizes of the major satellites of the solar system.*

deposit whose nature remains uncertain. Suggestions that it is due to material sputtered off from the darkish Phœbe do not seem at all convincing, and it is much more likely that the dark covering has welled up from inside Iapetus itself.

Titan, of course, was in a class of its own, and was not likely to have any bearing upon what would be found at Uranus: but there seemed every chance that the Uranian satellites would be of the same nature as the icy moons of Saturn. Craters, ridges and troughs were expected. But there was always the possibility of a major surprise – as had indeed been the case with both the other satellite systems; nobody had expected Europa or Enceladus to be so smooth, and few astronomers had anticipated the redness or volcanic nature of Io. So as Voyager 2 flew on toward Uranus, there was a keen sense of excitement from all those who were even marginally involved in the mission.

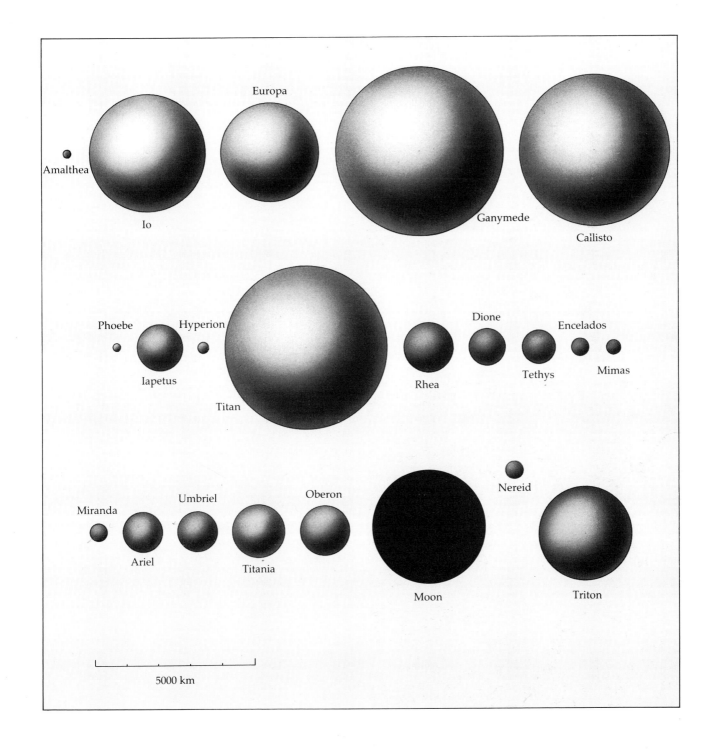

Approach to Uranus

Voyager 2, let us repeat is an 'old' probe. It had been launched well over eight years before its pass of Uranus, and technology had improved considerably since 1977. There had also been some alarms en route. On 5 April 1978 the CCS automatically switched to the back-up receiver. Unfortunately the back-up receiver had concealed a faulty capacitor, so that it was unable to lock on to the frequency of the transmitted signal. When the prime receiver was switched on, it then immediately failed. Consequently, almost the entire mission has been flown with the malfunctioning receiver, and the engineers have to tune in to it with the precision of a short-wave radio, since the exact frequency for communication is very sensitive to the changing temperature of the receiver. If the transmitted frequency were not within 96 hertz of the receiver rest frequency, Voyager 2 would turn a deaf ear to signals sent from Earth.

Then, almost a hundred minutes after the closest approach to Saturn in August 1981, the azimuth motion of the all-important scan platform, carrying the cameras, jammed, so that numerous unrepeatable observations of the planet, its satellites and the surrounding environment were lost. At first it was thought possible that the space-craft had been damaged by collision with an icy particle from Saturn's rings, but after careful tests (involving simulations of the Voyager which were as perfect as possible) it was decided that lubrication failure was the cause. Apparently the scan platform had seized up during a high-rate slew corresponding to 1 degree per second; the lubricant migrated away from a tiny shaft-gear interface spinning at 170 revolutions per minute, causing it to heat up and expand. Although it took two days to restore the scan platform motion, a detailed investigation of this serious problem was needed before the Uranus encounter. The 4.5-year cruise from Saturn to Uranus provided an excellent opportunity to service the space-craft. After an exhaustive number of

The path of Voyager 2 through the Uranian system.

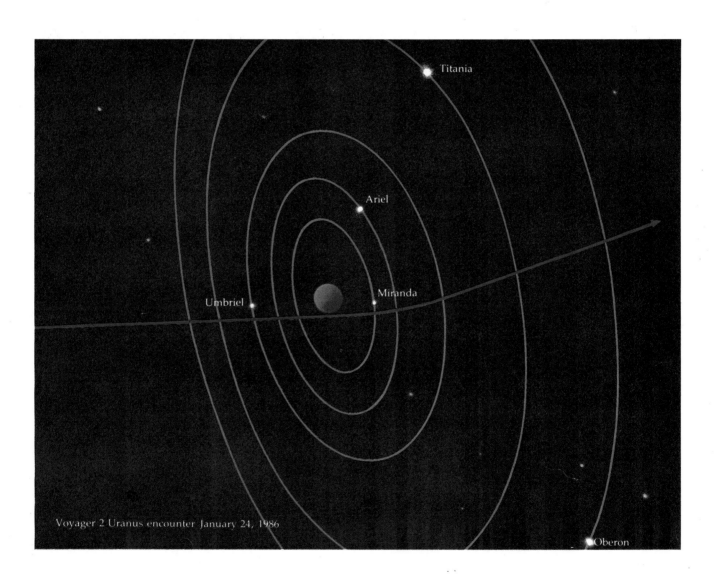

Titania

Ariel

Umbriel

Miranda

Voyager 2 Uranus encounter January 24, 1986

Oberon

tests, it was found that the actuator behaved correctly when operated at a low scan rate of only 0.08 degrees per second, and the observational sequences were planned accordingly. Thankfully, there were no problems with this mechanism at Uranus.

Since the sunlight at Uranus is only 1/400 as strong as at the Earth, it was obvious that very long exposures would be needed in order to obtain good images of the Uranian system. Some exposures were as long as 96 seconds, and these images are always vulnerable to smearing due to the motion of the space-craft. The Attitude Control System was altered to reduce the angular rates by a factor of 2 or 3 in order to give better compensation for the impulses generated by stopping and starting the tape recorder, thereby reducing any possible smear in the images. Furthermore, schemes had to be introduced to pan the cameras to compensate for the motion of the space-craft at the time of

closest approach to some of the targets. The clarity of the high-resolution images, particularly those of Miranda, gives a good indication of the success of these techniques.

The tremendous distance of Uranus from the Earth also creates problems by decreasing the strength of the radio signals. At Jupiter, the data rate was 115.2 kilobits per second, while at Uranus it was reduced to 21.6 kilobits per second. Special techniques were used to send back the imaging data in a more ecconomical manner, without reducing the scientific information content, to overcome the reduction in data rate. Instead of transmitting the full 8 bits (256 grey levels) of each picture (pixel), only the difference between the brightness of successive pixels is transmitted. This technique reduces the number of bits required for each image by 70 per cent.

The main tracking during the actual encounter was to be carried out from Australia. The great radio telescope at Parkes, 64 metres across, had been called in to link with the Deep Space Network 'dish' at Tidbin-

The 64 metre radio telescope at Parkes, NSW, Australia.

The NASA Deep Space Network (DSN) receiving station at Tidbinilla, near Canberra, Australia.

billa, near Canberra; every scrap of power was needed. The only hitch at ground level came at a critical moment in the Uranus mission. At the same time, the European space-probe Giotto was on its way to Halley's Comet, and it too was being tracked from Australia. When there was a sudden sharp drop in signal strength, the Giotto planners feared the worst, and called for help. At the Jet Propulsion Laboratory, it was agreed to take on the tracking of Giotto from Goldstone, in California, another of the Voyager 2 stations. The emergency was short-lived, and was found to have been due to the fact that part of an underground cable in Australia had been accidentally ploughed up by an innocent farmer. No data were lost, but the episode did stress the very close collaboration between all the space missions.

Much more serious was an earlier alarm, four days before the Uranus encounter. There was a memory failure in one of the space-craft's on-board computers, severely affecting a portion of the imaging data. At a distance of nearly 3000 million kilometres, when the one-way transmission time is 2 hours 45 minutes, it is a little difficult to carry out running repairs; but commands were sent, and the system was brought back into operation. Finally all went well, and we now have a wealth of spectacular observations which have changed our whole understanding of the Uranian system.

NASA's 64 metre dish at Goldstone, California, continues the world-wide Deep Space Network. The whole DSN is constantly being improved to gain information from the weakening transmissions of receding spacecraft.

The official encounter sequence began on 4 November 1985, when Voyager was still 103 000 000 kilometres from the planet. By 21 January, when scientists of all nations were assembling at the Jet Propulsion Laboratory, excitement was mounting. Several small satellites had already been discovered, all closer-in than Miranda, innermost of the previously-known moons, but outside the rings. Early on 21 January the space-craft was just over 4 000 000 kilometres from Uranus, and moving in fast; just over 3 000 000 kilometres early on 22 January, below 2 000 000 kilometres on the 23rd. Closest approach to Uranus took place at 17 hours 58 minutes 51 seconds GMT on 24 January. The distance from the centre of the planet was then 107 000 kilometres, and from the cloud-tops only 81.543 kilometres. Voyager 2 passed within 16 kilometres of the scheduled target point, and the error in timing was only 1 minute 10 seconds.

By then some fascinating data had been received. At first the planet appeared featureless; it was all very different from the spectacular, brilliantly-coloured pictures which Voyager 2 had sent back from Jupiter and Saturn at comparable range. But three days before encounter, a banded structure started to show up, though it looked unfamiliar because of the 'pole-on' view. A few definite clouds were seen, one of which appeared to have a rotation period of about 16.25 hours, but others were faster moving – in one case a period of only 15 hours was suggested; this particular cloud was at a latitude of 41 degrees north.

It was beginning to look as though there would be no radio emission, and therefore no appreciable magnetosphere, when the situation suddenly changed. Radio waves were picked up, and so was evidence of a magnetic field, which turned out to be intrinsically stronger than that of the Earth, though at the Uranian cloud-tops it is weaker. The main surprise was that the magnetic axis is inclined to the rotation axis by as much as 60 degrees, far more than with any other planet (with Saturn, as we have seen, the magnetic and rotational axes almost coincide). Voyager 2 went through the boundary of the magnetosphere eight hours before closest encounter, at a distance of around 470 000 kilometres from the centre of the globe.

There was, inevitably, considerable discussion about nomenclature. As the axial tilt is 98 degrees to the perpendicular, which is the true 'north pole' and which is the 'south pole'? Add the fact that whichever pole is regarded as 'north' is also magnetically 'south', and it is easy to understand why speakers at the daily press conferences became somewhat confused!

By 23 January data were arriving quickly – not only from Uranus itself but also from the satellites, which were much more interesting than anyone had expected. One important point concerned the relative abundance of the Uranian atmosphere of the two lightest elements, hydrogen and helium. Ground-based analyses had indicated that the helium content might be as much as 40 per cent, which would have been difficult to explain and which would have presented theorists with baffling problems. However, the Voyager analyses, carried out as the space-craft moved in, showed that the real helium abundance was of the order of 12–15 per cent, similar to those of Jupiter, Saturn and also the Sun.

Temperatures were measured. Surprisingly, it was found that the dark side of the planet was at a higher temperature than the daylight side, and there was virtually no difference between the poles and the equator. A high layer of haze – which could well be termed a photochemical smog – was found around the sunlit pole, and the day side was also found to emit large amounts of ultra-violet radiation, a phenomenon which was promptly christened the 'electroglow'. Something of the same kind had been recorded at both Jupiter and Saturn, but it was much more pronounced with Uranus, and may not be due simply to auroræ, as had been tacitly assumed. At present the electroglow remains rather a mystery.

Most of the vital information was expected to come during the day of the encounter itself, 24 January. The timetable was as follows:

EMT *(subtract 8 hours for PST)*

14.00	Voyager 230 000 kilometres or 10 Uranian radii from Uranus
15.08	Closest to Titania (365 000 kilometres)
16.12	Closest to Oberon (471 000 kilometres)
16.20	Closest to Ariel (130 000 kilometres)
17.03	Closest to Miranda (32 000 kilometres)
17.16	Ring-plane crossing
17.59	Closest to Uranus (81 543 kilometres)
20.38	Occultation of Voyager by Uranus (till 22.45)
20.45	Closest to Umbriel (325 000 kilometres)
24.00	Voyager now 650 000 kilometres from Uranus.

The main encounter was then over; by early on 26 January the space-probe had receded to more than 2 000 000 kilometres from Uranus, and more than 5 000 000 kilometres by the 29th. Data continued to be received; one of the last really spectacular pictures to be received was that of the whole ring-system, showing 'dust' spread through it and looking misleadingly like a picture of the ring-system of Saturn.

The official encounter mode ended on 25 February. Voyager was then put into what is termed a cruise mode, when most of the equipment is switched off; there is still a long way to go, and as soon as Uranus passed out of range the members of the team had already started to look forward to the pass of the most distant giant, Neptune.

Now let us look at some of the results obtained from Uranus by Voyager 2.

The magnetosphere of Uranus

The evidence for an Uranian magnetosphere, and its possible characteristics, have been debated for many years. The effect upon a magnetosphere of the strangely-tilted but rapidly-rotating planet was previously unknown. However, the likely existence of a magnetosphere was conjectured once the observations of the Lyman-α emission from Uranus were made from the very successful satellite known as the IUE (International Ultra-violet Explorer). Voyager 2 made it possible to study the precise nature of this region of charged particles surrounding the planet.

The magnetosphere is a *windsock*-shaped region surrounding the planet, with the head directed toward the Sun and the tail streaming out behind in the solar wind. The region is formed when energetic particles are trapped within a planet's field lines. Several planets had previously been found to possess magnetic fields and associated magnetospheres. Mercury's magnetic field is weak, despite its relatively massive iron-rich core, while Venus and Mars have virtually no magnetic fields at all; but the Earth, Jupiter and Saturn all have substantial magnetospheres, and the Jovian region is so large that if it could be seen visually it would appear to us to have the same angular diameter as the Sun. Since in composition and rapid rotation Uranus is not too unlike the other giant planets, it seems reasonable to invoke what might be termed a 'magnetic Bode's Law', which would clearly indicate the presence of a strong Uranian magnetic field and extensive magnetosphere.

The first direct evidence of the Uranian magnetic field was drawn from the radio emissions, five days before Voyager's closest approach, when the spacecraft was still at a distance of 27.5 Uranian radii (27.5 R_U from the planet. Voyager subsequently crossed the bow shock at 23.5 R_U. (This region marks the boundary of the solar wind and the planetary magnetosphere, creating a shock-front which resembles that caused by an ocean liner on the surface of the sea.) Finally, at a distance of 18 R_U, Voyager entered the magnetosphere, revealing a strange magnetic dipole field with an axis tilted at almost 60 degrees to the rotational axis of the planet, and offset by 0.3 R_U from the centre of the globe. The dipole moment is 0.23 gauss per R_U^3, which corresponds to a surface magnetic field in the range of 0.1 – 1.1 gauss. The period of rotation of the magnetic field, as determined from the periodicity of the radio emissions, is 17.24$\pm$ 0.01 hours; now, at last, we have determined the length of the Uranian rotation. This corresponds to the rotational period of the deep interior of the planet, and therefore of the solid body of Uranus, so that all the meteorological wind observations are measured relative to this rotation period.

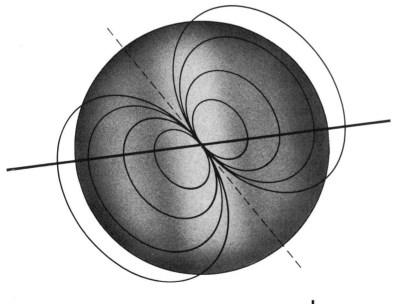

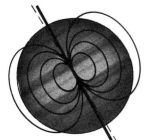

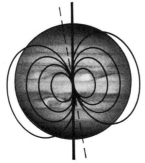

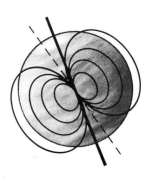

This figure compares the structure of the magnetospheres of Uranus with Jupiter, Saturn and the Earth and it clearly shows that the magnetic and rotational axes are not generally aligned in the same direction. Uranus top, *has the largest offset between these axes, while the Earth* left, *Jupiter* centre *and Saturn* right *have varying differences in the directions of these axes. The magnetic field lines in each case, extend into space and form a protective cage around the planet, trapping charged particles and sweeping them into space as the planet rotates.*

The magnetic field, whose north pole is currently on the sunlit side of the planet, is created by a dynamo action deep in Uranus' interior. The location of the dynamo region may be found from the requirement of the presence of an electrically conducting fluid at this level. The conductivity in the upper hydrogen layer must be too small to create the dynamo, but the necessary physical conditions are found at the intermediate depth in the oceanic layer, where the high temperatures and pressures in the huge water ocean are sufficiently electrically conducting to generate the observed field.

The magnetic field of Uranus is certainly very strange. It is the most inclined magnetic field of any object yet studied in the Solar System, as Table 3 shows. In an astrophysical context, it may be described

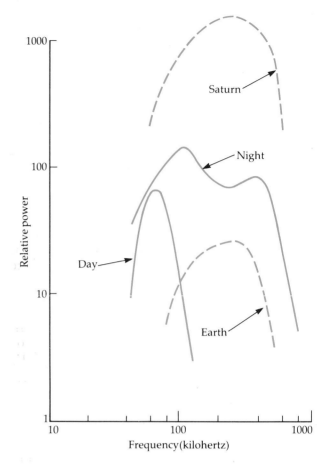

Right *The radio spectrum observed Radio Astronomy experiment for both the inbound (day) and outbound (night) portions of the Voyager 2 path through the Uranian system. The spectrum is compared with the corresponding observations for the Earth and Saturn. A relative power of 1 corresponds to a flux density of 100 Jy at a distance of 1 AU.*

Below *The field lines of the Uranus magnetic field compared with the rotational axis of the planet.*

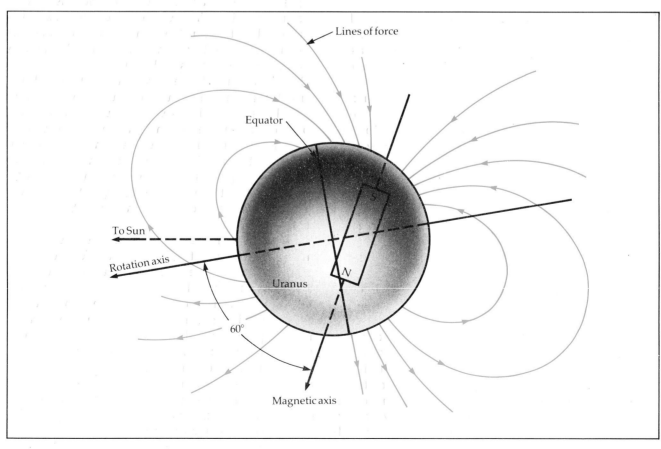

as an oblique rotator. There is also the possibility that we are observing a reversal in the polarity of the field, which would account for the large inclination; in this case the Uranian magnetic field might be in a 'flip-flop' motion. This situation is not unusual. The Earth's field has reversed at least nine times in the past 3 500 000 years, and metal particles in the rocks on the sea-bed have been found to be pointing in different directions which correspond to different magnetic epochs. Alternatively, the magnetic offset could be the result of a massive collision with an object about the size of the Earth, which could also have tipped the planet over, though – as we have seen – there are problems about this idea. However, it does seem that there may have been many collisions in this distant part of the Solar System, and the remarkable landscape of Miranda, to be described later, provides supporting evidence.

Table 3 *Some Magnetic Properties of the Planets*

Planet	Tilt, degrees	Dipole equatorial magnetic field, gauss	Size of magnetosphere, planetary radii
Mercury	14.0	0.0033	1.4
Earth	11.7	0.31	10.4
Jupiter	−9.6	4.28	65
Saturn	0.0	0.21	20
Uranus	−60.0	0.23	18

The dipole field is deformed by interaction with the streaming solar wind, which produces a magnetotail similar to that of the Earth. The Uranian tail has a radius of 42 R_U at a distance of 67 R_U behind the planet, where the plasma sheet is about 10 R_U thick.

The magnetosphere contains an extensive distribution of charged particles, composed mainly of electrons and hydrogen ions. The plasma region is composed of two principal ion populations: the warm region (10 electron volts) inside about 7 R_U, with a maximum density of 2 ions per cubic centimetre (2 cm^{-3}) inside the orbit of Miranda; the hot component (1000 electron volts) is confined to the region outside 5 R_U. Although the temperature in this region exceeds 500 000 000 degrees, the density of the material is so low that any exposed flesh would freeze immediately! The major source of material in the plasma may be the ionization of the extended hydrogen corona, but the ionosphere and the solar wind may also make contributions.

The Voyager results suggest that the energy density of the plasma is surprisingly small compared with that of the magnetic field. The orientation of the rotational axis also has a significant effect upon the characteristics of the region; this is because the radial convection of the plasma is fundamentally different than with any other planetary magnetosphere, in which the radial transport is normally dominated by diffusion. However, rapid convective transport on a time-scale of about 40 hours may prevent the accumulation of a significant density of heavy ions sputtered from the surfaces of the satellites.

Both the ring-system and the satellites lie within this hostile charged-particle environment of the magnetosphere. This is not a new situation. With both Jupiter and Saturn we have seen the important interactions between the charged particles and the ring/satellite systems as the embedded particles are swept away. However, the differences between each planetary system will mean variations in the individual characteristics of these intractive regions. Certainly the Uranian system has the darkest satellites and ring known at present, with the albedoes of some bodies as low as 3–5 per cent.

Beyond the orbit of Miranda, the trapped ion population is dominated by protons. However, the proton fluxes are too small to distort significantly the magnetic field, though they may be enough to alter the chemistry of the surfaces of the satellites and to create the unusual dark surfaces in a period of less than about 100 000 years.

Energetic electrons, with temperatures above 20 kiloelectron volts, are observed throughout the magnetosphere. They end abruptly at about 18 R_U on the dark side of the planet, where the magnetotail begins. There is an intense 'whistler-hiss', with 'chorus' emissions, inside a distance of 8 R_U, and it is possible that these radio emissions cause electron precipitation, thereby contributing to the auroræ detected on the planet's night side.

Both Miranda and Ariel absorb the protons and electrons strongly, and this must affect the surface chemistry of the satellites themselves through local heating. However, these interactions change over periods of time. The extreme tilt of the Uranian magnetic pole means that the satellites will sweep over a broad range of magnetic latitudes as the planet spins, and this in turn means that there will be a strong latitudinal variation in the trapped radiation field around the satellites and rings.

Uranus is so far away that it is not possible to detect any radio emissions from Earth – unlike the case of Jupiter, where, as we have seen, radio waves were picked up more than thirty years ago. The radio emissions from Uranus show a left-handed polarization, and appeared to reach maximum when the negative magnetic pole, currently on the dark side of the planet, was tipped toward Voyager 2. Observations from the day side give the peak frequency as about 60 kilohertz, while the dark side emissions extend up to 800 kilohertz. This suggests that there is a large diurnal asymmetry in the plasmasphere which may be affecting the propagation of the radio waves.

The rings

As we have seen, it was Taylor's occultation technique which provided the first evidence of the ring-system of Uranus. The rings known before the Voyager 2 missions were known as 6, 5, 4, α, β, η, γ, δ, and ϵ. Earth-based observations indicated that the rings ranged in width from about 100 kilometres at the widest part of the ϵ ring to only a few kilometres for most of the others. Further Earth-based observations suggested that the reflectivity of the entire ring-system was only from 1.6 to 2 per cent, and placed an upper limit of 0.0015 on the optical depth of the ring-system. If the diffuse material of the rings is distributed uniformly throughout the system, this would suggest that the ring thickness is about 0.2 kilometre.

The Voyager 2 encounter offered an unique opportunity for studying the ring-system – a procedure which will not, unfortunately, be repeated in the near

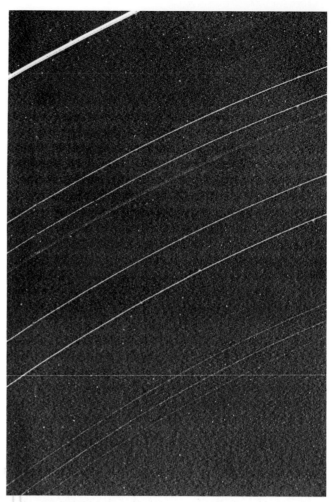

Right *The rings of Uranus, which barely show the new ring feature 1986 U1R between the ϵ ring (outermost) and the δ ring, the next bright ring.*

Below *The resolved structure of the ϵ ring.*

Opposite *Picture of the rings of Uranus showing the broad diffuse component of the η ring.*

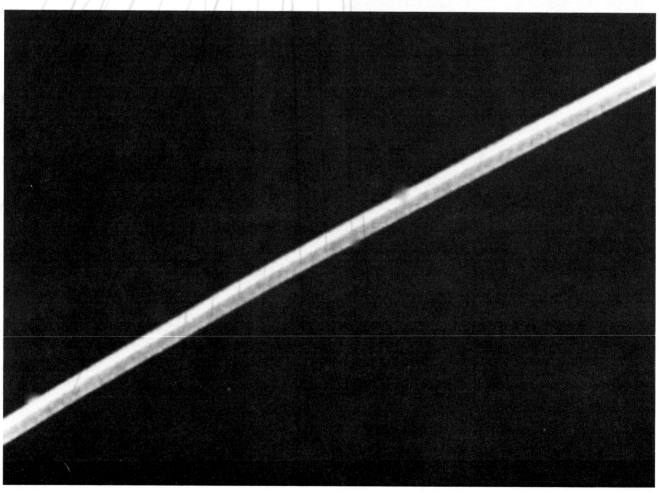

future, at least so far as we can tell. Moreover, the space-craft trajectory meant that the rings could be observed in transmitted light. In addition, occultations of the stars Nunki (Sigma Sagittarii) and Algol (Beta Persei) were observed with the photopolarimeter and ultra-violet spectrometer, thereby providing invaluable information about the structure of the rings.

The new rings were found in addition to those previously known, whose distances from Uranus ranged between 41,880 kilometres and 51,190 kilometres. Of the new rings, one (provisionally known as 1986 U1R) lay between the ϵ and δ rings, and the other (1986 U2R) closer-in than ring 6. However, the observations made in backscattered light show a multitude of new rings, so that the total number is probably uncountable. There are also five new ring arcs outside the ϵ ring. The term 'arc' is used at this stage, since it is not known whether the features are continuous. Details of the rings are given in the Table 4.

Above *Uranus and the outermost ϵ ring seen by Voyager 2 on the 28 November, 1985 from a distance of 72.3 million kilometres. The picture is a computer composite of six images which are then equivalent to an exposure of 84.5 seconds.*

Opposite *Observations of the newly-discovered rings and ring arcs by the Voyager 2 photopolarimeter instrument from occultations by σ Sagittarii and β Persei.*

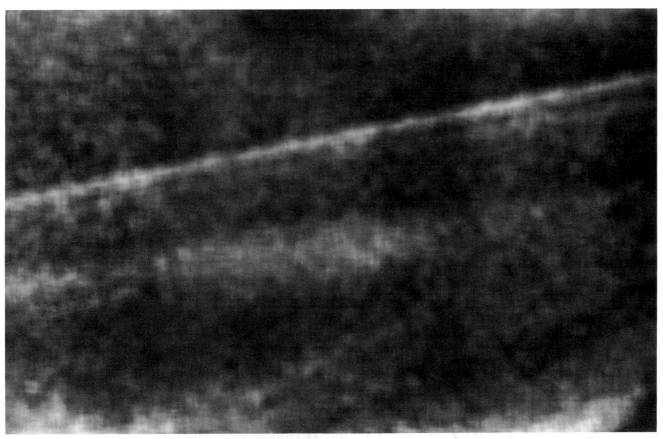

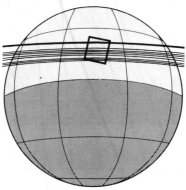

The viewing geometry for the image shown at the left. The black square corresponds to the size and orientation of the image shown.

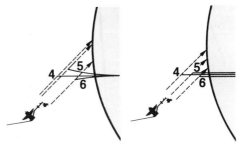

The spacecraft views of the rings 4, 5 and 6 during the observational sequence. The configuration of the ring geometry matches the predictions for ring positions made from Earth based models.

Table 4 *Properties of the rings of Uranus*

Ring	Distance, 10^3 kilometres	Eccen-tricity	Inclin-ation, degrees	Width, kilo-metres	Optical depth
1986 U2R	37.00–39.50	0.0	?	2500	0.0001–0.001
6	41.85	1.0	0.066	1–3	0.2
5	42.24	1.9	0.050	2–3	0.5
4	42.58	1.1	0.022	2	0.3
α	44.73	0.8	0.017	8–11	0.3–0.4
β	45.67	0.4	0.006	7–11	0.2
η	47.18	0.0	close to zero	2	0.1–0.4
γ	47.63	0.0	0.006	1–4	1.3–2.3
δ	48.31	0.0	0.012	3–9	0.3–1.0
1986 U1R	50.04	0.0	?	1–2	0.1
ε	51.16	7.9	close to zero	22–93	0.5–2.1

Shepherd satellites in the Uranian rings. The two satellites Cordelia and Ophelia are on either side of the ε ring.

It is perhaps surprising that only one pair of 'shepherd satellites' has been found; these satellites are Cordelia and Ophelia, adjacent to the ε ring. Originally we had expected to find at least eighteen shepherd satellites spanning the then-known rings. The outer edge of the ε ring is sharp, and this position corresponds closely to a resonance of the satellite Ophelia, while the inner edge is correspondingly close to a resonance of Cordelia. These resonances do not overlap. Consequently, the interaction in this region must be similar to that found between Mimas and the outer edge of Saturn's B ring, but it differs from the shepherd satellites and the narrow F ring of the Saturnian system. The rings of Uranus are thin and narrow, so that there must be some mechanism to keep them in position; perhaps there are very small dark satellites with albedoes of from 1 to 2 per cent, and diameters below 10 kilometres, yet to be discovered – or there could be weak resonances between the satellites. Certainly the charcoal-black material of the ring-system does make the detection of very small objects extremely difficult.

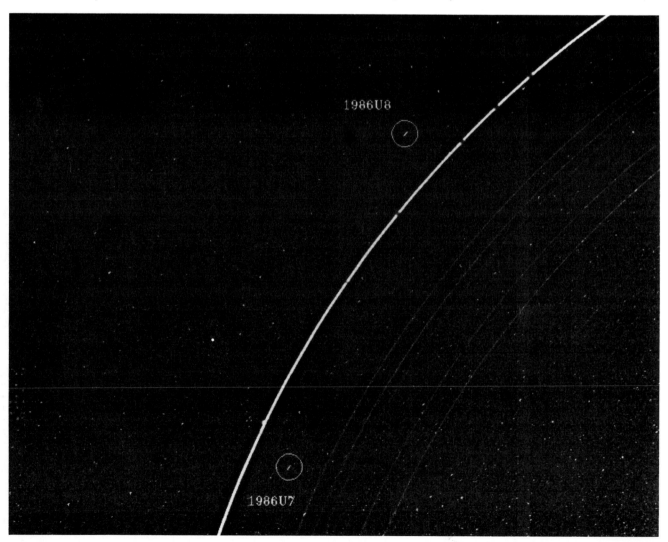

1986U8

1986U7

A further major surprise concerns the sizes of the ring particles. The ϵ ring appears to be composed of dark boulders about 1 metre in diameter, and is therfore quite unlike the rings of Saturn; indeed, the sharp edge of the ring suggests that its thickness is less than 150 metres and that the particles are actually below 30 metres across. The apparent absence of smaller particles in this portion of the ring-system may be a result of the atmospheric drag from the extended hydrogen atmosphere in which the rings are embedded. This mechanism limits the lifetimes of micron- and centimetre-sized particles to periods in the range of from 100 to 10 million years. The ϵ ring is dark, with an albedo of only 5 per cent; it appears to be grey in colour, and there does not seem to be any significant colour difference between any of the ring elements.

There may be some micron-sized objects inside the location of the new ring 1986 U1R to at least the outer edge of 1986 U2R. The dust distribution is highly structured, and this region at least seems to resemble what is often called the D ring of Saturn (though the D ring is not visible from Earth, and is not so well defined as the other Saturnian rings). In some regions the dust opacity is several orders of magnitude greater than anything found in Jupiter's ring, or the G and E rings of Saturn.

When Voyager 2 passed through the ring-plane, it received a significant number of impacts in a 400-kilometre thick band lying at about 116 000 kilometres from the centre of Uranus. The plasma wave instrument recorded a maximum impact rate of 20–30 hits per second, which, for micron-sized particles, indicates a number density of about 0.001 to the cubic metre.

Several of the main rings show considerable longitudinal variability in their opacity. Both the δ and γ rings vary by more than a factor of 2, while the ϵ ring appears to disappear completely in some places. This clearly indicates that the rings of Uranus are relatively young and dynamic, and were probably formed after the planet itself.

When Voyager 2 passes by the rings of Uranus, it captured images of an almost uncountable number of rings as a result of solar backlighting effects.

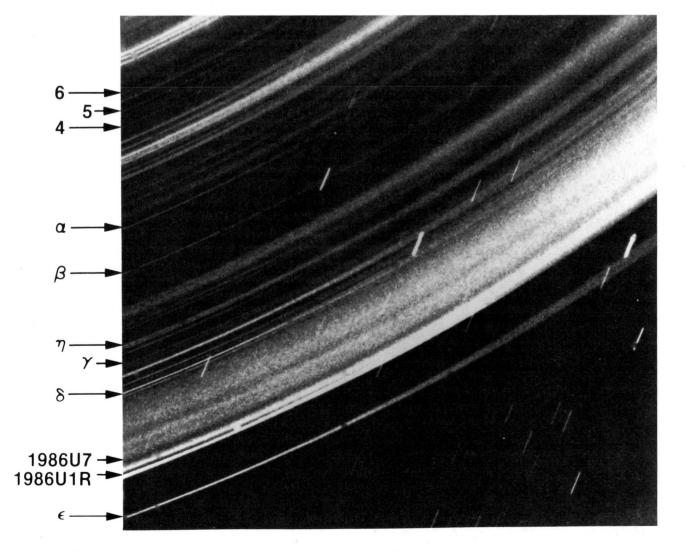

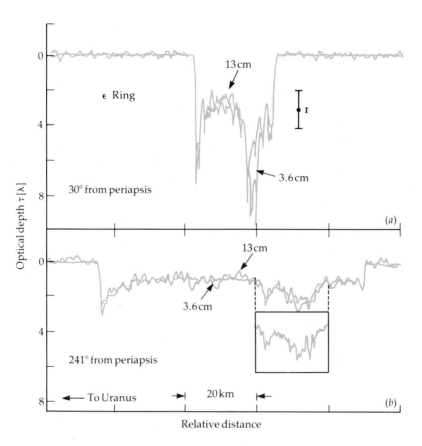

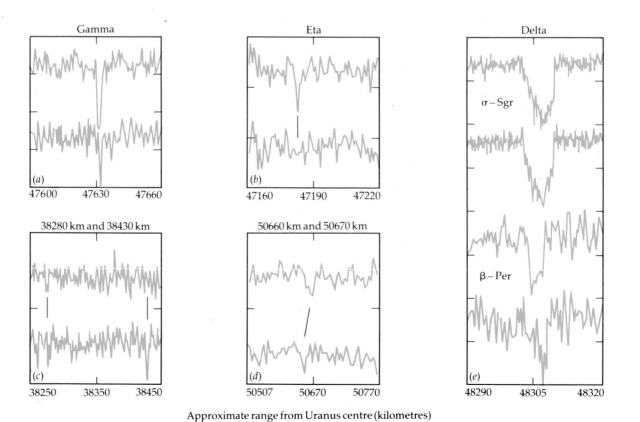

Above *Voyager 2 observations of the optical depth variations of the ε ring at 3.6 and 13 cms showing considerable structure in the ring, which is evident in the visible images, for example page 60.*

Below *The temperature structure of the atmosphere of Uranus at the north pole, south pole and the equator observed by the Voyager 2 IRIS instrument.*

Approximate range from Uranus centre (kilometres)

The structure of Uranus

Uranus has a diameter of 52 400 kilometres, more than four times that of the Earth, but it has only 15 times more mass than the Earth and a mere 5 per cent of the mass of Jupiter. The density of Uranus is 1.2 grams per cubic centimetre (that is to say, 1.2 times as dense as water) which is greater than with the larger gas-giants, Jupiter and Saturn.

Uranus is made up principally of hydrogen and helium, in approximately solar proportions, with small amounts of methane, ammonia and their photo-chemical products. There are also heavier materials such as oxygen, nitrogen, carbon, silicon and iron. Beneath the extensive cloud-layers there is believed to be an ocean of superheated water overlying the Earth-sized rocky core. It has been suggested that this exten-sive ocean, more than 8000 kilometres deep, may originate from collisions with thousands of millions of comets during the early history of the Solar System; comets are, after all, composed primarily of watery substances, and are found in the outer regions of the System. The numerous collisions and the high pres-sure of the region where the water is now located would account for its superheated state.

Uranus orbits the Sun in a period of 84 years; as we have seen, the colour is greenish, and the rotation period is 17 hours 14 minutes. The brightness temper-ature is about 55 K.[†] However, unlike the other giant planets, Uranus does not seem to have an internal heat-source. Certainly both Jupiter and Saturn have such sources; so presumably has Neptune, since its effective temperature of 57 K is about the same as that of Uranus even though Neptune is so much further away from the Sun.

According to recent data, about 30 per cent of the heat radiated by Uranus may come from the interior. This compares with around 70 per cent for Jupiter and Saturn, and indicates that Uranus must have lost most of the internal heat which it presumably had when it was formed.

The three-layer model of the internal structure – rocky core overlaid by a liquid ocean which is in turn overlaid by the hydrogen-helium rich atmosphere – has also been challenged. On an alternative model, there is still a rocky core, but this is overlaid by a dense atmosphere in which the gases and ices are mixed. What we actually see, of course, is the upper cloud-layer, which is mainly methane. Uranus is so cold that methane can condense at a high level, and the icy methane clouds hide the clouds of ammonia and water beneath. Above lies a thin upper atmosphere, made chiefly of hydrogen together with some helium and neon.

Atmosphere

The atmospheres of the giant planets are hydrogen-rich reducing envelopes, in contrast to the oxidizing atmospheres of the terrestrial planets. The spectrum of Uranus is dominated by strong hydrogen and methane absorptions, particularly in the red and infra-red parts of the spectrum. It is this methane absorption at red wavelengths which is largely responsible for the greenish colour of the planet's disk.

Helium was detected during the Voyager 2 observa-tions, and, as we have noted, its mass fraction is about 0.27 ($\pm$ 0.08). This is greater than the equivalent values for Jupiter and Saturn, but the helium mass fraction for the solar nebula is also about 0.28, and therefore the same as for Uranus. Very few other constituents have yet been detected in the atmosphere. Traces of deuterium have been found in the form of HD and CH_3D. Ammonia has been found only in the deeper layers of the atmosphere, since the temperature is too low for ammonia to exist in the neighbourhood of the cloud-tops (the temperature there is only about 64 K). However, microwave measurements suggest that there is less ammonia at these levels than might be expected in a mixing ratio based upon solar abun-dances, as we find with Jupiter and Saturn. Curiously, the entire microwave region of the spectrum, centred upon a wavelength of 1 centimetre, seems to have a temperature of 50 K, which is higher than the conden-sation temperature of ammonia; therefore, gaseous ammonia, in equilibrium pressure with solid ammonia, is not characteristic of the entire planet. It is difficult to explain why Uranus is in this state. The depletion of ammonia may be due to the formation of ammonium hydrosulphide clouds at a lower level in the atmosphere, but this would require the atmos-phere to have a sulphur/nitrogen abundance greater than the Sun. Carbon may be more abundant than in the Sun, which is logical if Uranus formed from an ice-rich material.

There may also be changes in the atmospheric struc-ture taking place at these deeper levels. Monitoring the microwave emission since 1965 has shown some surprising variations. The secular increase in tempera-ture during the past fifteen years has now been reversed. The temperature maps at 6 centimetres show an increase in magnitude from equator and pole, with a warm ring surrounding the south pole of the planet; this is, of course, the pole which was in sun-light at the time of the Voyager 2 encounter. The spatial variability of the brightness distribution may be responsible for the unexpected variability of the micro-wave spectrum.

[†] 55 K gives the temperature on the Kelvin scale, where zero is taken as absolute zero – equivalent to −273 degrees Centigrade.

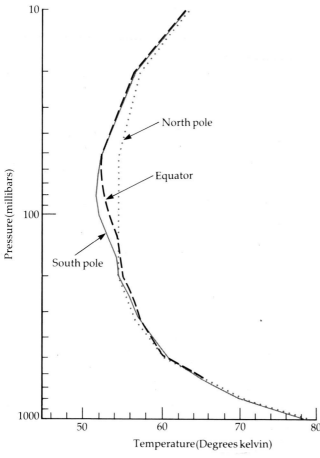

Atmospheric structure

Although there is a general similarity in the composition of the atmospheres of the giant planets, there are important differences between Uranus and the other three. Temperature profiles are determined mainly by heat balance under the influence of solar and internal heating, and it is here that Uranus is unusual; as is apparent from its effective temperature of about 57 K, it has a negligibly small internal heat source. When compared with Neptune, we find that, surprisingly, the more distant planet has the stronger stratospheric inversion. While the dominating absorption of sunlight by methane is the principal process in each case. Neptune also has layers of methane clouds and photochemical hazes which add further to the atmospheric heating at these levels; Uranus, on the other hand, seems to have an atmosphere which is relatively clear. In the troposphere, the structure tends to follow the adiabatic lapse-rate in the usual way.

Left *The variation of atmospheric pressure as a function of temperature for the polar and equatorial regions is clearly seen.*

Below *The variation of temperature as a function of latitude between the equator and the sunward pole.*

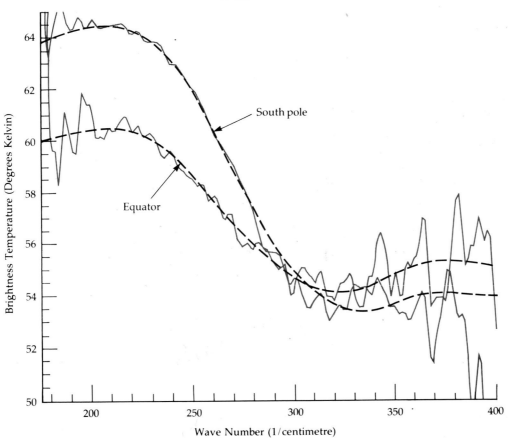

The axial inclination of 98 degrees has some interesting seasonal effects. Each pole spends 21 years in constant sunlight, while the equatorial region has a double season. The Voyager observations showed some effects of the seasonal lag in the temperatures in the atmosphere. In the lower stratosphere, at pressure levels between 200 and 100 millibars, the dark pole was 2–3 degrees warmer than the illuminated pole; the seasonal variation at these levels is estimated to be about 5 K. Evidently the atmosphere in the winter hemisphere is slowly cooling to its equilibrium state, but since the radiation relaxation time for the atmosphere of Uranus is some 700 years, thermal changes are extremely slow.

There are some spatial variations in the atmospheric temperatures in the neighbourhood of the cloud-tops corresponding to the 800 millibar level. While there is a negligible pole-to-equator gradient at this level, where the temperature is about 64 K, there is a small feature in the region of latitude 10–40 degrees south where there is a minimum temperature of 63 K at latitude 30 degrees. It is surprising that the temperature appears to drop in this region, where there is less overlying haze. This dip in the temperature is absent from the dark winter hemisphere.

One of the possible models of Uranus' interior composition. In this model, due to Hubbard and MacFarlane, the rocky core amounts to just less than a quarter of the total mass of the planet. Next comes a composite water, ammonia and methane ice mantle with a mass of around 9.5 Earths (this accounts for some 65% of the total mass of Uranus). The remaining 11%, or so, of the planet's mass is in the hydrogen – helium atmosphere.

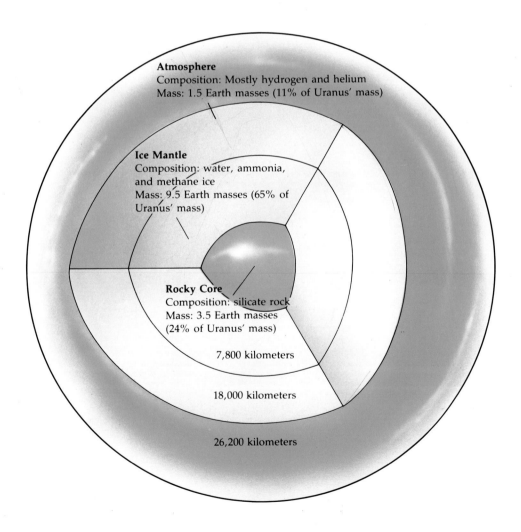

Atmosphere
Composition: Mostly hydrogen and helium
Mass: 1.5 Earth masses (11% of Uranus' mass)

Ice Mantle
Composition: water, ammonia, and methane ice
Mass: 9.5 Earth masses (65% of Uranus' mass)

Rocky Core
Composition: silicate rock
Mass: 3.5 Earth masses (24% of Uranus' mass)

7,800 kilometers

18,000 kilometers

26,200 kilometers

Clouds, hazes and atmospheric motions

Uranus is the first planet so far from the Earth that no dynamical activity shown by clouds can be detected by ground-based observations. We had thought the atmosphere to be cold and clear to great depths, with little evidence of many discrete cloud features similar to those of Jupiter or Saturn. The brightness variations at pressure levels less than 1 bar on Uranus are due mainly to the Rayleigh scattering of sunlight by hydrogen and helium molecules and the scattering and absorption of this radiation by the photochemically produced hazes. Therefore, the Voyager encounter was the first opportunity to investigate the Uranian weather system. However, the task was much more difficult than might be thought. The mission was not optimized for Uranus; at the time when the space-craft was launched, in 1977, we would have been more than satisfied with good results from Jupiter and Saturn, so that Uranus was sheer bonus – apart from the added complication that the observations were made with ageing instruments which do not have the best spectral sensitivity. However, we have made some amazing discoveries.

Below The appearance of Uranus seen by Voyager at violet *left,* orange *centre, and methane 0.619 micrometres wavelength* right, *for a phase angle of 17° and a resolution of 300 km per line pair. The contrast in each of the processed images has been enhanced to bring out the detail in the atmosphere. The concentric banding is centre on the pole of rotation and not on the subsolar point. The sequence of images shows the wavelength dependence of the aerosol and cloud features. The aerosol particles dominate the violet image and hid the cloud features which become more evident at the longer wavelengths, particularly in the methane.*

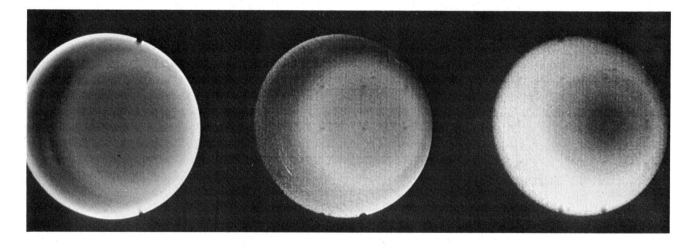

As Voyager 2 sped away from Uranus after its historic encounter, it returned this image of the planet that could never be seen from the Earth. At the time this image was recorded, the spacecraft was some one million kilometres away from Uranus.

Before the Voyager mission, CCD images by Smith and Terrile at the deep methane band centred at 0.890 micrometres had shown some structure in the extended haze layers, together with evidence of a bright limb, but the wavelength which they used was beyond the sensitivity of the Voyager cameras, so we still had no real idea of any possible cloud features.

The appearance of Uranus at violet, orange and methane wavelengths corresponding to 0.426, 0.605

A time sequence of four orange images showing the movement of small scale features in the southern hemisphere of Uranus between latitudes of 25° and 40° S. The image sequence shows a clear indication of anti-cyclonic motion in the atmosphere.

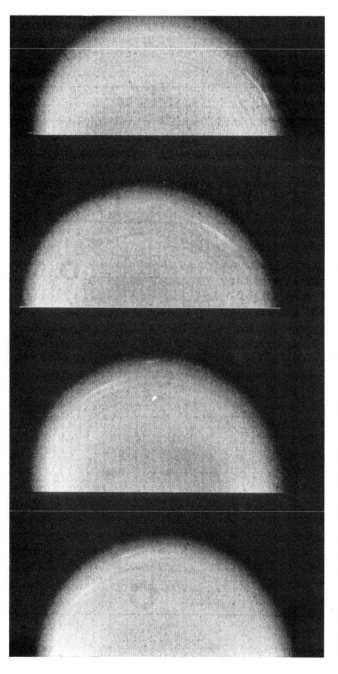

and 0.618 micrometres respectively shows definite structure in the clouds. In violet light, the planet displays a banded appearance about the pole of rotation, with the darker bands being located at higher latitudes. In orange light, the banded appearance is different; now the planet is darkest in the middle and low latitudes, while a bright band is evident at 50 degrees. Finally, at the methane wavelength, the research provides some information about the structure of the clouds, since high clouds will appear bright on the image while low clouds will look dark. The mid-latitude band, where in the orange image a thin cloud feature is evident in the south-east quadrant of the picture, is a region lacking in the usual overlying haze layers. These high-level hazes are thought to be created photochemically, and to be made up of acetylene and ethane particles.

Therefore Uranus does have a banded appearance, and the cloud motions indicate a predominately zonal circulation where the winds are blowing in an east–west direction rather than north–south. This circulation resembles the flows on Jupiter and Saturn and, to a lesser extent, the motions in every planet in the Solar System which has an appreciable atmosphere. All planetary weather systems are zonal, in spite of the substantial differences in solar heating distribution, so that it must be the rotation of the planet which organizes its weather system.

It is in the 20–45 degree latitude band of Uranus, where sunlight can penetrate to the warmer levels, that a few discrete cloud systems have been found. This has provided the first direct evidence of local motions in the Uranian atmosphere. These clouds, whose tops lie at a level of 1–2 bars, are probably composed of methane particles, and morphologically resemble the convective plume clouds seen in the equatorial region of Jupiter. This is consistent with the signature of a methane ice cloud in the region of 900–1300 millibars, identified in the radio occultation data. At 50 degrees south, the methane clouds appear as bands with a latitude scale of 700 kilometres, which suggests discrete sources for the clouds and little horizontal diffusion. All the clouds in this region move with prograde winds in a westerly jet, with speeds of 40–160 metres per second faster than the planet's rotation period of 17.24 hours. The orientation of the plume tails and the evidence of the positive direction of the winds relative to Uranus' interior suggest that the zonal velocity increases with altitude in the region of from 27 to 70 degrees south. Surprisingly, this conclusion is in contrast to the implication which may be drawn from the thermal winds expected from the observed latitudinal temperature gradient. We can then only assume that dynamical processes are creating a more complicated situation than current theories can explain.

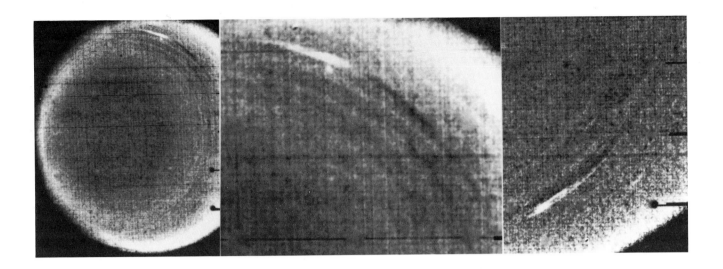

Above *The brightness temperature spectrum of the atmosphere of Uranus at the south pole and the equator.*

Below *The measured zonal velocities and rotational periods in the southern hemisphere of Uranus derived from racking the displacements of the observed clouds, see for example opposite.*

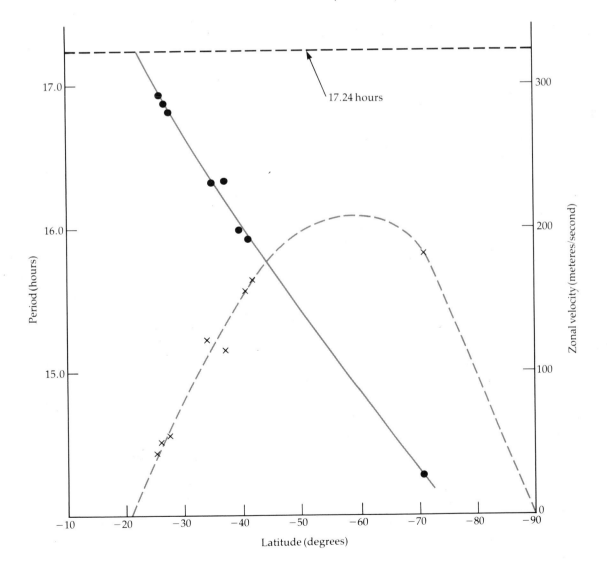

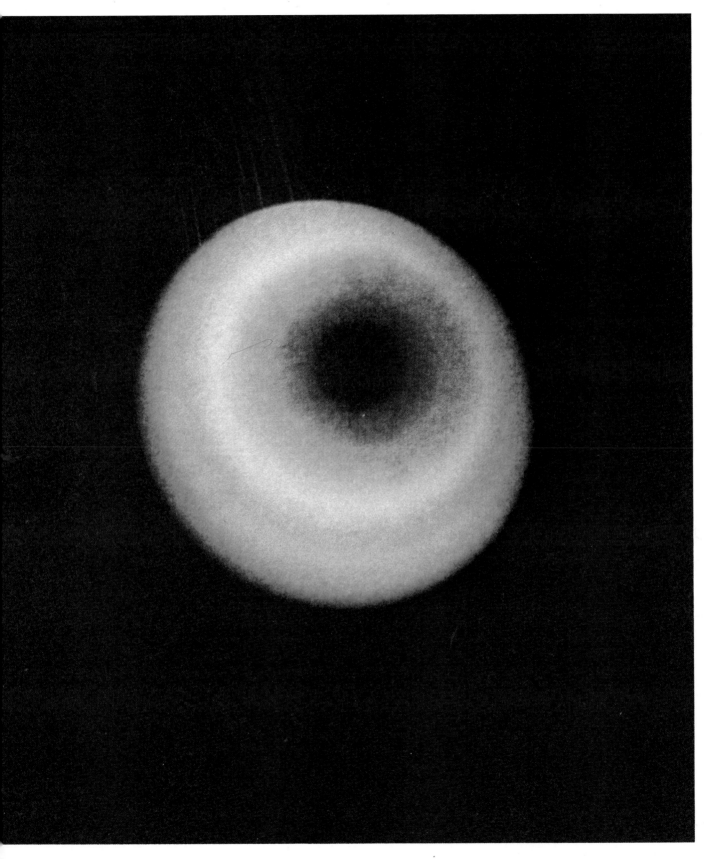

The **left** *picture shows the colour picture of Uranus as would be seen by the human eye. The* **right** *picture shows the appearance of the planet using a composite of ultra-violet, violet and green images which emphasises the structure in the haze layers in the planet's atmosphere. A thick haze layer can be seen over the south pole of the planet.*

Extended atmosphere and magnetospheric interactions

As we have seen, the huge 'electroglow', which spreads out for some 50 000 kilometres (twice the radius of the planet) is something of a mystery. Unlike the aurora, it is seen only over the sunlit hemisphere. There are suggestions that it is due to the collision of electrons with hydrogen molecules, the problem here being that the Sun's energy seems to be too weak to accelerate the electrons sufficiently. However, excited by low-energy electrons, the emission occurs well above the homopause, and consists primarily of Lyman-α. The electroglow process, which may be driven in some way by the coupling of atmospheric winds into the ionosphere, also produces 10^{29} atoms per second through the dissociation of hydrogen. About half of the atomic hydrogen escapes, and forms the hydrogen corona, which extends out to the rings with a density of 100 atoms per cubic centimetre at the ϵ ring.

The aurorae observed on the dark side of Uranus are produced by the excitation of the molecular hydrogen by 10-kiloelectron volt electrons, resulting in an auroral oval of about 15–20 degrees centred on the magnetic pole.

A further unexpected observation has been made in the region of the thermosphere, where on the sunlit side the temperature is 750 K, while on the dark side it is 1000 K. Perhaps some dynamical processes are responsible for this surprising result.

Ionosphere

Above the mesosphere comes the ionosphere, where the density is very low and the electrical conductivity increases with altitude. The name is derived from the higher proportion of atoms and molecules which are ionized.

The Voyager 2 observations at 3.6 and 13 centimetre wavelengths provided the first evidence of the Uranian ionosphere in the region between 2 and 7 degrees south. They show clear signatures of well-separated, narrow ionospheric layers at altitudes of 2000–3500 kilometres above the millibar pressure level. Sharp layers of this type have been detected for Jupiter and Saturn, and may be similar to the sporadic E layers on Earth. It is possible that the extended Uranian ionosphere may extend up to 10 000 kilometres or even more.

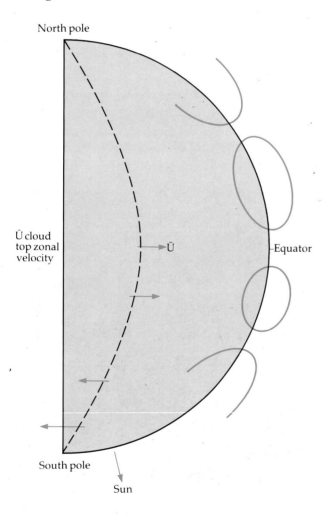

A sketch of the main solar driven weather systems in the atmosphere of Uranus. The main Hadley circulation comprises the two cells adjacent to the equator.

The satellite system

Before the Voyager 2 mission, five satellites of Uranus were known. As we have seen, two were found by Herschel, two by Lassell and one, much later, by Kuiper. They are not easy objects. One of the present authors (PM) has often observed Titania and Oberon with his 39-centimetre reflector, but Ariel has only been glimpsed with extreme difficulty, and Umbriel not all, while Miranda is far beyond the reach of such a telescope.

Ten further satellites were discovered from Voyager; all are very small, and are surprisingly dark. One (Puck, the first to be found) was surveyed from reasonably close range; not surprisingly it proved to be cratered, as indeed are all the five satellites previously known. The densities of the main satellites, and very probably of the new ones also, are low, so that they may well be composed of water-ice together with some rocky material and a little methane clathrate. However, the proportion of methane-ice must be very low, since the density of satellites made up mainly of frozen methane would be only about 0.53 that of water – and the satellites are considerably denser than that. The darkness of the surfaces may therefore be due to the presence of carbonaceous material, or to the bombardment by charged particles in Uranus' magnetosphere, which would cause heating of the surfaces with consequent alterations in their chemistry.

Clearly the Uranian system is different from those of the other giant planets, and is as interesting as any, as the Voyager pictures have shown. Data for the satellites are as shown in Table 5. All these data have been revised to some extent as a result of the Voyager mission. In particular Umbriel now seems to be slightly larger than Ariel instead of appreciably smaller – because its surface has a lower reflecting power. No doubt further small satellites exist, but in any case Uranus is well supplied with attendants.

Table 5. *Properties of the satellites of Uranus*

Satellite	Mean distance from Uranus, kilometres	Period, days, hours, minutes	Diameter, kilometres
Cordelia	49 771	7 55	15
Ophelia	53 796	8 55	20
Bianca	59 173	10 23	20
Cressida	61 777	11 07	70
Desdemona	62 676	11 24	50
Juliet	64 352	11 50	70
Portia	66 085	12 19	90
Rosalind	69 942	13 24	50
Belinda	75 258	14 56	50
Puck	86 000	18 17	170
Miranda	129 900	1 9 50	484±10
Ariel	190 900	2 12 29	1160±10
Umbriel	266 000	4 3 28	1190±10
Titania	436 300	8 16 56	1610±10
Oberon	583 400	13 11 07	1550±20

The family portrait of the five large satellites of Uranus seen by Voyager 2 on the 20 January, 1986 from distances ranging from 5 to 6.1 million miles.

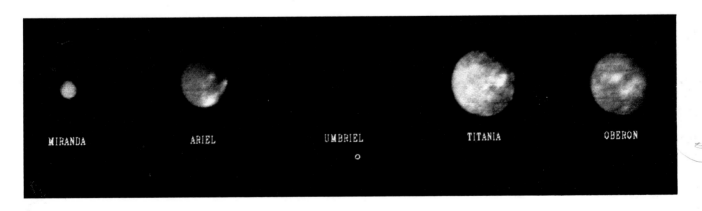

MIRANDA   ARIEL   UMBRIEL   TITANIA   OBERON

The larger satellites

Long before the flight of Voyager 2, the movements of the five known satellites had been worked out very precisely. Large telescopes had no trouble in following them. When a pole of Uranus is facing the Sun (as at present) the orbits of the satellites will appear circular; when the equator is presented (as happened in 1966) the orbits will carry the satellites 'up and down' – a situation unlike any other in the Solar System, and yet another consequence of Uranus' extraordinary axial tilt.

The names for the first four seem to have been suggested by Sir John Herschel, son of Sir William, though it may be that Lassell had used them earlier. Titania and Oberon, of course, are the characters from Shakespeare's *A Midsummer Night's Dream*. Ariel comes from *The Tempest*, but also from Pope's *The Rape of the Lock*, which also gives Umbriel as 'a dusky sprite' – and from an early stage Umbriel was known to be fainter, and therefore 'duskier', than Ariel, Titania or Oberon. Certainly Gerard Kuiper named Miranda, in 1948, on the assumption that the other names were Shakespearian; Miranda also comes from *The Tempest*. All the names of the new satellites are from Shakespeare, except Belinda, who is the heroine in Pope's poem in which Umbriel appears. Puck, of course, is one of Shakespeare's creations in *Midsummer Night's Dream*.

Additional data for the main satellites, revised in view of the Voyager 2 results, are shown in Table 6. Before the Voyager mission our knowledge even of these satellites was very fragmentary. It was usually thought that Titania was the brightest of them, but in 1950 W. H. Steavenson, one-time President of the Royal Astronomical Society and known to be an observer of exceptional skill, made a series of measurements with his 76-centimetre reflector, and gave the magnitudes as 13.9 for Ariel, 14.0 for Titania, 14.1 for Oberon and 14.8 for Umbriel, indicating that Ariel was the brightest of the four and presumably the largest. He also found that the magnitudes of Titania and Oberon vary over a small range (around 0.2) and that these variations were independent of the position in orbit. He concluded that Titania and Oberon revolved round Uranus with their axes more or less in their orbital planes, as Uranus does with respect to the Sun, so that we would not see them pole-on. But nobody

really knew, and it was also almost impossible to find out anything about their surfaces.

Surface details were quite beyond the range of any Earth-based telescopes, but spectroscopic observations indicated that Ariel, Titania and Oberon showed clear evidence of water-ice. So did Umbriel, but there were significant differences, indicating that Umbriel might in some ways be 'the odd one out' in the Uranian system. Because it was the faintest of the four main satellites it was also assumed to be the smallest.

There seemed every reason to expect that the satellites would be essentially similar in nature to the icy satellites of Saturn. Titania and Oberon were thought to be about the size of Rhea, while Ariel was taken to be roughly the equal of Dione. As Voyager 2 moved in towards its fly-by, one of the leading planetary geologists at the Jet Propulsion Laboratory, Laurence Soderblom, said that he anticipated 'impact craters and little else'. This sounded reasonable enough, but very soon the satellites began to reveal unexpected characteristics.

Oberon was the first, early on the day of encounter, 24 January. There were craters and bright rays, but also darkish patches on the crater floors. Next came Titania; craters again, but also rift valleys hundreds of kilometres in length. Ariel was even more spectacular, with Titania-like rifts together with broad, smooth valleys; the surface looked 'younger' than those of the outer pair. Umbriel did indeed prove to be different. It showed a darkish, relatively bland surface, with some craters, but with a general impression of being extremely ancient. There was also one feature, not well shown because of foreshortening, which was bright; was it a crater – or something else? It was compared with a doughnut!

Not all four satellites were equally well imaged, because of Voyager's trajectory; Ariel was approached almost four times more closely than Oberon. But it was with Miranda that the main surprises came. The incredibly varied landscape caused universal astonishment. There were grooves, craters, valleys, towering ice mountains, a 'chevron' and a feature which was nicknamed 'the race-track'; it was said that Miranda's surface was the most fantastic ever found, not even excluding the huge volcanoes of Mars.

What do these results tell us about the origin of the satellites? Unfortunately, not a great deal. The fact that they move virtually in the plane of the planet's equator, and that they are moderately large, argues against any idea that they might be captured asteroids. If it is true that Uranus was 'tipped over' by a tremendous impact during or soon after its formation from the solar nebula, we must assume that the satellites were created later. It has even been suggested that after the catastrophe, Uranus captured a body with about 5–10 per cent of the mass of Uranus itself, after which the

Table 6 *Physical properties of the satellites of Uranus*

Satellite	Orbital eccentricity	Orbital inclination, degrees	Mean opposition magnitude	Density (water = 1)	Albedo
Miranda	0.017	3.4	16.5	1.26±0.39	0.34±0.02
Ariel	0.0028	0.0	14.4	1.65±0. 3	0.40±0.02
Umbriel	0.0035	0.0	15.3	1.44±0.28	0.19±0.01
Titania	0.0024	0.0	14.0	1.59±0.09	0.28±0.02
Oberon	0.0007	0.0	14.2	1.5 ±0.1	0.24±0.1

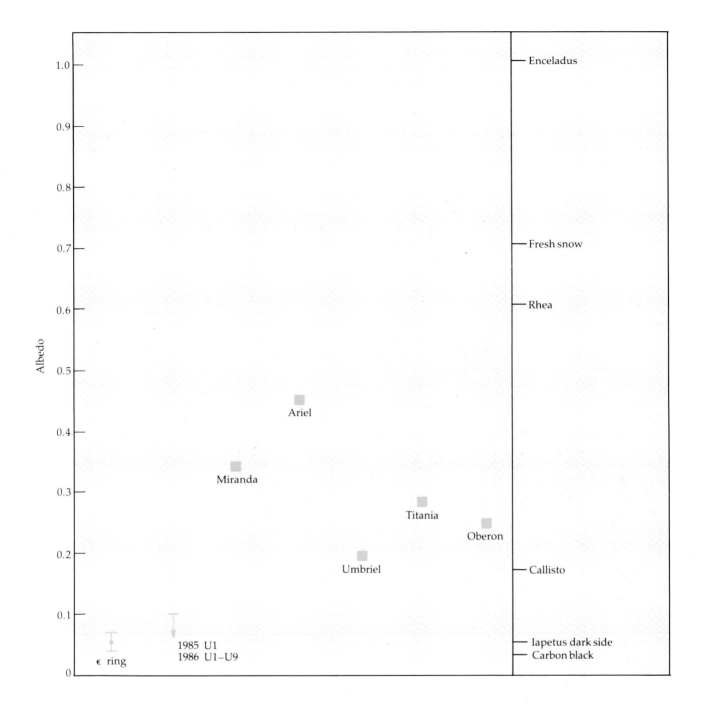

captured body broke up to produce the satellites. But there is a great deal to be learned. With Jupiter's main satellites, it was once said that 'there is no such thing as an uninteresting Galilean'. It is equally true that there is no such thing as an uninteresting satellite of Uranus.

The variations of the albedos of the satellites of Uranus compared with values for some of the satellites of Saturn, Jupiter and other well-known light and dark materials.

Oberon

Oberon was the first-discovered of the satellites of Uranus; Herschel found it shortly before he had his first glimpse of Titania. Oberon was also the first of the satellites to be reasonably well imaged from Voyager 2; the best picture of it was taken on 'encounter day' from 660 000 kilometres, and has a resolution on Oberon's surface of about a dozen kilometres.

Even earlier, it had become evident that Oberon has a brownish surface pitted with craters. Though the average albedo is low, there are several large craters such as Othello, which are the centres of bright ray systems, not unlike those on Jupiter's outermost Galilean, Callisto. But there is a major difference between Oberon and Callisto (quite apart from the fact that Oberon is much the smaller). Callisto is saturated with craters, and the surface seems to indicate that nothing much has happened there since an early stage in the story of the Solar System, but inside the craters of Oberon there is strange dark material, so that something must have flowed out from below the surface.

The instinctive answer is that after the formation of the craters, however this may have happened, 'dirty water' was sent out through cracks in the crust, filling the floors of some of the craters. There is one very prominent formation Hamlet near the centre of the disk in the best Voyager pictures, which shows this very well. We may be dealing with what may be termed 'icy vulcanism', though naturally there is no suggestion that there have ever been true volcanoes on Oberon.

There may be a parallel with Saturn's satellite Iapetus, which is of much the same size, and is icy. As we have seen, a substantial part of one hemisphere is covered with dark material which is very probably of internal origin – though whether it is similar to the dark material inside the craters of Oberon remains to be seen.

Another feature of Oberon is what appears to be a lofty mountain, some six kilometres high, shown on the best Voyager picture exactly at the edge of the disk, so that it protrudes from the limb (otherwise it might not be identifiable). Whether or not it is exceptional is something else which we do not know.

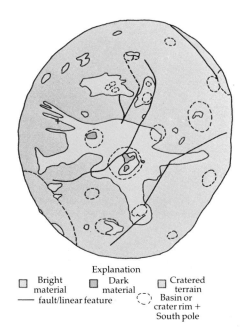

Explanation

Bright material — Dark material — Cratered terrain — fault/linear feature — Basin or crater rim + South pole

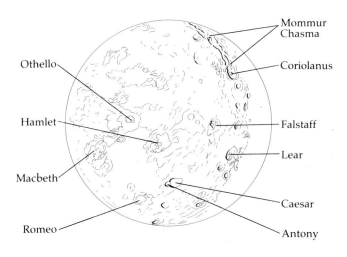

Othello, Hamlet, Macbeth, Romeo, Mommur Chasma, Coriolanus, Falstaff, Lear, Caesar, Antony

Top right The outermost moon, Oberon, seen from a distance of 2.77 million miles by Voyager 2 on the 22 January, 1986.

Titania

During its flight through the Uranian system, Voyager 2 actually passed Titania first, just over an hour before the closest approach to Oberon. But the resolution with Titania was decidedly better. The best picture was obtained at a range of 369 000 kilometres.

Titania is only marginally larger than Oberon, but is somewhat more massive (if represented by snooker balls, it would be hard for even a skilled player to tell which was which). Like Oberon, Titania has an icy, cratered surface, but there are some significant differences, notably the lack of dark material inside Titania's craters.

The images indicate that there has been considerable activity at some stage during the satellite's history – more than with Oberon. On the best Voyager view, there is a 200-kilometre crater, Ursula, on the lower part of the disk, and a 300-kilometre crater, Lucetta, near the top. Even more spectacular are the linear troughs or fault valleys, which are up to 1500 kilometres long and as much as 75 kilometres wide in places. The faulting does not seem to be oriented entirely at random. There are at least two systems (possibly even reminiscent of the grid system on our Moon?) and one suggestion is that the features were produced when the interior of Titania expanded, breaking the crust and forming the valleys and troughs. Whether or not all tectonic activity there has come to an end is not certain, but at least the surface seems less ancient than Oberon's. One trench-like feature, Messina Chasmata, was imaged as it lay on the terminator, or boundary between the illuminated and dark hemispheres, so that was excellently shown.

Like Oberon. Titania has craters which are the centres of bright ray systems. There is, too, one large formation, Gertrude, which may be more in the nature of a basin than a crater; it was well shown on a picture taken from 500 000 kilometres, with a resolution down to below 10 kilometres. Moreover, the existence of relatively young 'frost deposits', sent out from the interior, is not impossible.

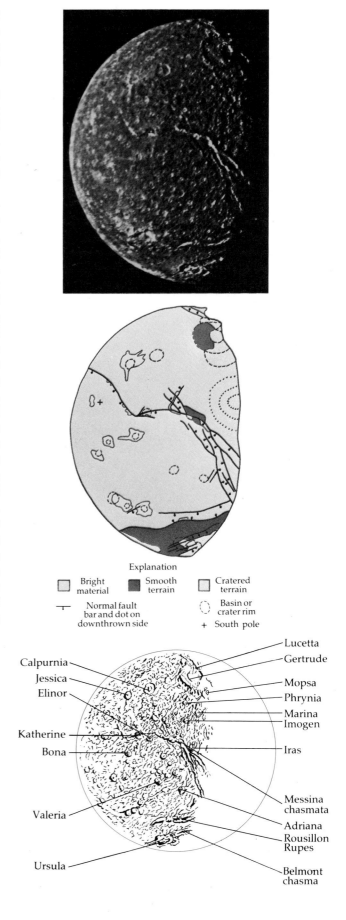

Explanation

▢ Bright material	◼ Smooth terrain	▢ Cratered terrain
⊢ Normal fault bar and dot on downthrown side		◌ Basin or crater rim
		+ South pole

Calpurnia
Jessica
Elinor
Katherine
Bona
Valeria
Ursula

Lucetta
Gertrude
Mopsa
Phrynia
Marina
Imogen
Iras
Messina chasmata
Adriana
Rousillon Rupes
Belmont chasma

Umbriel

Umbriel, so appropriately named after the 'dusky sprite', is very different from the other principal satellites. This had been suspected well before the Voyager encounter, and it was amply confirmed. Umbriel's surface is dark, with an albedo of less than 20 per cent, and there is a relatively small range in the brightness of the various features, so that even from close range the surface appears rather monotonous. There are craters, but most of them are subdued, and there are no bright ray systems.

The most detailed picture, taken from a range of 537 000 kilometres, has a resolution of about 10 kilometres. One very prominent crater, Skynd, is shown on the terminator; it is at least 110 kilometres in diameter, with a bright central or near-central peak, so that it is one of the few albedo features on Umbriel which really stand out. The other bright feature, Wunda, is much more problematical. It appears near the top of the picture, at approximately longitude 270 degrees and close to the equator (remembering that Umbriel, like Uranus itself, was being observed almost pole-on). The feature appears to be a ring about 140 kilometres across, but it is so badly foreshortened that its form cannot be made out, and we cannot even be sure that it is a crater at all; it could be a frosty deposit associated with a crater wall, though nothing comparable has been found elsewhere on the visible disk of Umbriel.

The problem of why Umbriel has its dark, subdued surface, and why it is unlike the other satellites, is not easy to solve. It would be more logical to expect that the level of past tectonic activity would increase with decreased distance from Uranus – Titania less inert than Oberon, and so on; certainly Ariel's surface appears to have been much more active in the past. Not so with Umbriel. (It is interesting, but almost certainly not significant, that with all the giant planets it is the second of the 'long-known' satellites which is exceptional; Europa in Jupiter's system, Enceladus in Saturn's and Umbriel in that of Uranus.)

Is Umbriel's icy surface coated with dark, powdery material? That the globe is essentially icy we cannot doubt; the overall density is slightly less than with the other main satellites. Alternatively, the water-ice crust might contain frozen methane which would be broken down by charged particles in the Uranian magnetosphere, leaving a layer of darker carbon; but again there seems no reason why Umbriel alone should be so affected, and there is no sign of methane-ice on the other satellites. The same objection applies to the idea that the whole globe could be a well-mixed combination of ice and carbonized material. The 'doughnut' feature Wunda makes matters even less easy to explain; if it and the central peak of the large crater Skynd are the only bright objects, it is not clear how they can have been produced. Certainly Umbriel is a puzzle.

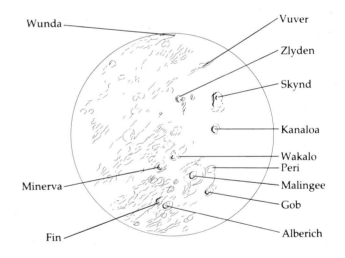

Ariel

While Umbriel has been inert since an early stage in its history Ariel has been active. Of this there can be no doubt. The two are almost equal in size; it is now believed that Ariel is very slightly the smaller, because Umbriel has proved to be darker and therefore larger than expected, but once more we are dealing with two bodies whose dimensions are virtually identical. In size, the Ariel/Umbriel pair is as similar as the Titania/Oberon pair; but just as Titania is rather denser and more massive than Oberon, so Ariel is denser and more massive than Umbriel.

The best Voyager picture of Ariel was taken from a range of 130 000 kilometres, and gives resolution down to 2.4 kilometres – which seems incredible when we remember how far away from Earth it is, and that before the Voyager mission it had never been seen except as a tiny disk too small for its diameter to be properly measured.

Voyager showed that there are plenty of craters, both bright-rimmed and ray-centres, but the overall aspect is by no means the same as that of Titania, and evidently the tectonic activity was spread over a long period. The dominant features are broad, branching, smooth-floored valleys. Some of these valleys, such as Korrechan and Sylph, are partly filled with obviously younger deposits which are less heavily cratered; there are also grooves, sinuous scarps and faults. It is likely that the valleys have formed over down-dropped fault blocks known as graben, presumably because of the expansion and consequent deformation of Ariel's crust.

It has been suggested that there may be an analogy with Enceladus, in Saturn's system, where there are smooth areas presumably due to fresh water-ice sent out from below the crust. But with Enceladus there is a possible resonance with the more massive satellite Dione, which, so to speak, has churned Enceladus' interior. There is no resonance between Ariel and any other satellite, and in any case it is not likely that any effects of this sort could have been strong enough to produce the violent activity which must have occurred on Ariel. Were the valleys produced by flowing water – or by glaciers? Or could some other liquid have been responsible? Even liquid carbon monoxide has been suggested (it is true that at these low temperatures, carbon monoxide would liquefy), but again there seems no clear reason why this kind of activity should be confined to Ariel.

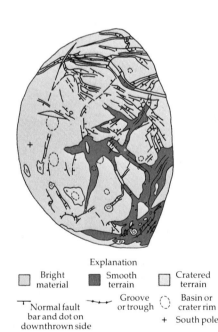

Explanation

Bright material | Smooth terrain | Cratered terrain
Normal fault bar and dot on downthrown side | Groove or trough | Basin or crater rim | South pole

Miranda

Miranda, the smallest of the previously-known satellites of Uranus, is also the most remarkable. In fact, it may be said to be the most bizarre world in the entire Solar System. Voyager 2 passed it at a distance of only 3 000 kilometres, and the pictures sent back gave a resolution down to 600 metres, so that the views of Miranda are actually more detailed than those of any other world except those upon which space-craft have landed.

'You name it, Miranda has it!' This was the comment made during a broadcast by one of the present authors (P.M.) from Mission Control on 25 January, just after the pictures had been transmitted. Miranda, no more than 484 kilometres in diameter, is a bewildering hybrid. There are features of all types, fairly obviously formed at very different epochs; almost everything is there, all concentrated in a small area.

In one image, three types of terrain are evident. One area is apparently ancient and cratered, consisting of subdued, rolling hills and worn-away craters of medium size. Then there is a grooved terrain, with linear valleys and ridges; next to it we find a complex region in which intersecting curvilinear ridges and troughs are abruptly stopped where the grooves begin. There is one large, dark, rectangular feature, Inverness Corona, which in another age would have suggested an artificial origin; it was nicknamed 'the chevron'. Near the limb is another huge, grooved area, Arden Corona, which has been called 'the race-track'. Together with scattered craters, fault valleys, parallel ridges, graben up to 15 kilometres across, scarps, fractures and troughs, Miranda presents an incredible picture.

How can these diverse features have been produced? One idea is that at an early stage in its development Miranda was hit by a large body and literally shattered, subsequently reforming; if the original composition had been a mixture of ice and rock, some of the reassembled fragments might have been purely icy and others purely rocky. But this would involve considerable heating, and in view of Miranda's small size this sounds improbable. Moreover, there are not many comparatively large craters.

It is often tacitly assumed that the craters on the satellites in the Solar System, including our Moon, are of impact origin. But not all astronomers agree, and the strange surface of Miranda may well indicate that the full story is highly complex. At any rate, it is safe to say that Miranda came as a surprise to everyone. It is a pity that we have been able to map only part of it; the night side was inaccessible to Voyager, but we have enough to guide us, and the analysis of all the results from Miranda will take a very long time indeed.

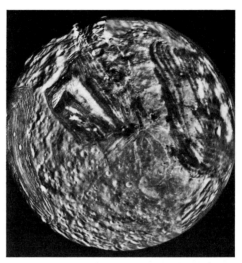

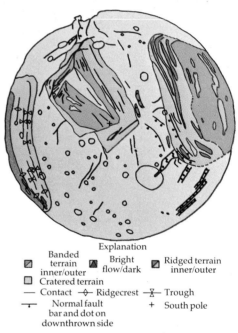

Explanation

Banded terrain inner/outer — Bright flow/dark — Ridged terrain inner/outer — Cratered terrain — Contact — Ridgecrest — Trough — Normal fault bar and dot on downthrown side — South pole

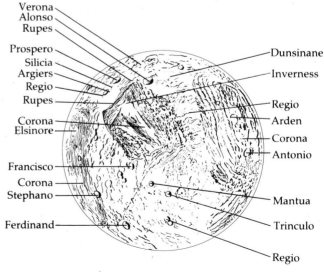

Verona
Alonso
Rupes
Prospero
Silicia
Argiers
Regio
Rupes
Corona
Elsinore
Francisco
Corona
Stephano
Ferdinand

Dunsinane
Inverness
Regio
Arden
Corona
Antonio
Mantua
Trinculo
Regio

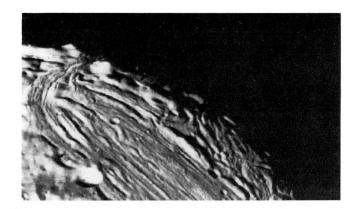 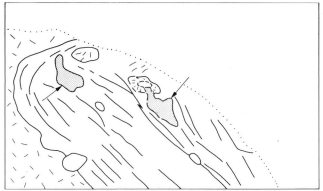

A region about 250 kilometres across on the satellite Miranda seen by Voyager 2 on the 24 January, 1986. Numerous craters are seen on the rugged, higher terrain suggesting that it is older than the lower regions of the moon. There are several scarps and possibly faults cutting the surface of the satellite. The impact crater is about 25 kilometres across.

The new satellites

One of the very first revelations from Voyager 2 came as early as 30 December 1985, almost a month before the closest approach. A new satellite was discovered. Its orbit lay closer-in than that of Miranda, though well outside the ring-system. It was given the provisional designation of 1985 U1, and was found to be very small, with a diameter of no more than 170 kilometres – less than half that of Miranda. There would have been no hope of detecting it with Earth-based telescopes. It has now been named Puck.

Fortunately, the trajectory of the space-craft showed that on 24 January, 'encounter day', Voyager would pass within 500 000 kilometres of the new satellite. Hasty calculations were made. The chance of obtaining a picture was much too good to be missed, and all went well. True, the image is of lower quality than those of the large satellites, but there were problems about obtaining it at all. The picture was recorded on the space-craft tape, and first played back during the late afternoon on 24 January (PST). An antenna-pointing problem at one of the Australian tracking stations led to a loss of data, and the image had to be transmitted a second time; it was finally received on 26 January, and was well worth waiting for. The satellite turned out to be almost spherical, with the very low albedo of only 7 per cent (not unlike that of our Moon). There was one large crater, another of fair size, and indications of many more. The picture gave a resolution of about 10 kilometres, which under the circumstances was remarkably satisfactory.

If there were one small inner satellite, there might be more – and indeed there were. A second (Portia) was found on 3 January, and others followed until the grand total of all known satellites had grown to fifteen. As was to be expected, the new bodies were small, with low albedoes. Luckily, Voyager was able to follow them for long enough to enable astronomers to work out their orbits.

All the newcomers except the two innermost are beyond the ring-system, but Cordelia and Ophelia are involved in it. Before the Voyager pass, nine rings were known, of which the two outermost were the δ and ϵ rings at distances of 48 300 and 51 200 kilometres respectively. On 23 January a Voyager picture showed a tenth ring, midway between δ and ϵ, at a distance from Uranus of around 50 000 kilometres. This placed it at very much the same distance as Cordelia, and it had already become apparent that Cordelia and Ophelia were 'shepherd satellites' of the ϵ ring.

This was not unexpected. Saturn's F ring has two shepherds, the satellites Prometheus and Pandora, neither of which is as much as 150 kilometres in diameter; Prometheus orbits just inside the F ring, Pandora just outside it. If a ring particle strays away from its orbit, it is 'herded back' by one or other of the shepherd satellites, and the F ring is kept intact. In the same way, Cordelia and Ophelia guard the ϵ ring of Uranus.

Because of the thinness and darkness of Uranus' rings, it had been suggested that there might be numbers of embedded shepherd satellites, but apparently this is not so (neither, for that matter, is it the case with Saturn's rings). Had there been any more shepherds, Voyager 2 would probably have found them. One the other hand, it is quite likely that there are several more minor satellites between the ring edge and the orbit of Miranda; ring they must be very small indeed – certainly not as much as 100 kilometres across – and Voyager could easily have missed them.

The nature of the small inner satellites is presumably bound up with that of the ring-system, though Puck may be more akin to Miranda. All move in the same plane as the rings, and their orbital eccentricities are very low. It is tantalizing that we will not be able to record them again for a long time in the future, but at least we have the satisfaction of knowing that they are there.

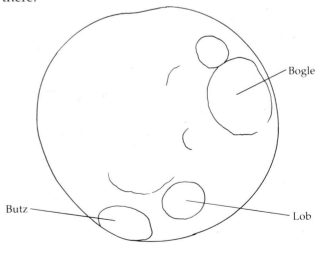

The newly-discovered satallite, Puck.

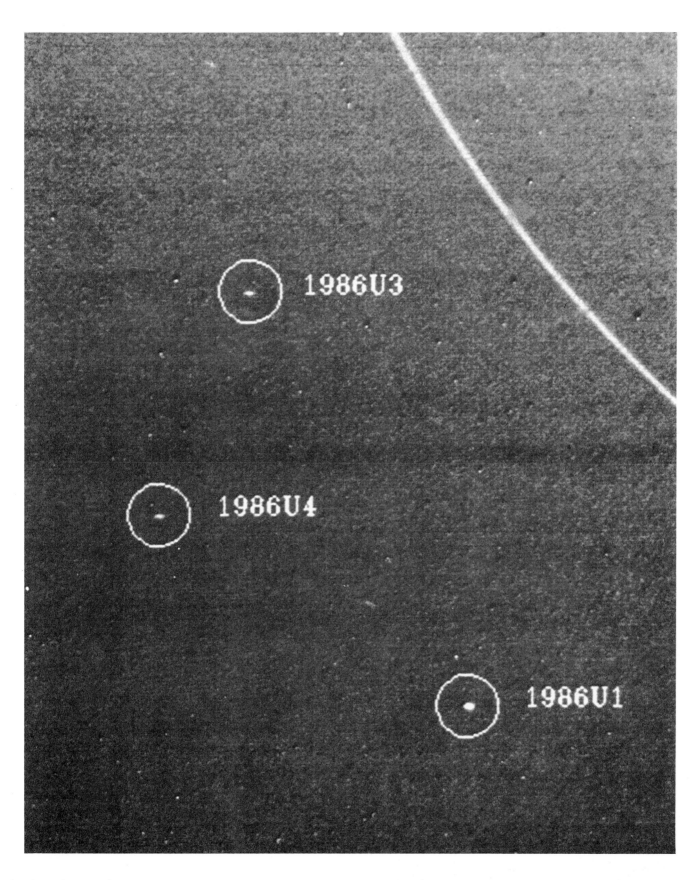

*Three of the newly-discovered satellites, Portia (1985 U1),
Cressida (1986 U3) and Rosalind (1986 U4), observed beyond the
ε ring when the space-craft was 7.7 million kilometres from the planet.*

Beyond Uranus

So far, Uranus is the most distant object to have been explored by a space-craft. But as we have seen, Voyager 2 still has work to do. It is on its way to a rendezvous with the outermost giant, Neptune, in August 1989.

The discovery of Neptune was due directly to measurements of the wanderings of Uranus. Alexis Bouvard produced a set of tables which seemed to be satisfactory, but after a while it became clear that Uranus was not behaving as expected. Bouvard therefore rejected all the old observations – that is to say, those made before 1781 – on the grounds that they were unreliable. However, before long the discrepancies showed up again. Bouvard, in the preface to his tables, had wondered whether Uranus might be being affected by some 'extraneous and unknown influence', but the first definite suggestion that there might be a still more remote planet seems to have come in

1834 from an amateur, the Rev. T. J. Hussey. Hussey wrote to one of Britain's leading astronomers, Dr (later Sir George) Airy, who was then at Cambridge, but who went to Greenwich as Astronomer Royal in 1835 on the resignation of John Pond, and remained in office until 1881. Hussey asked whether the motions of Uranus would have any clue as to the position of a new planet, but Airy's reply was not encouraging. Later, in 1837, Airy received a letter from Eugène Bouvard, nephew of Alexis, who pointed out that there was now a very marked error between the actual and predicted positions of Uranus, and went on: 'Does this suggest an unknown perturbation caused to the motion of this planet by a body situated further away? It is my uncle's idea.'

Influence of Neptune upon Uranus. Before 1822 Neptune was accelerating Uranus; after 1822, retarding it.

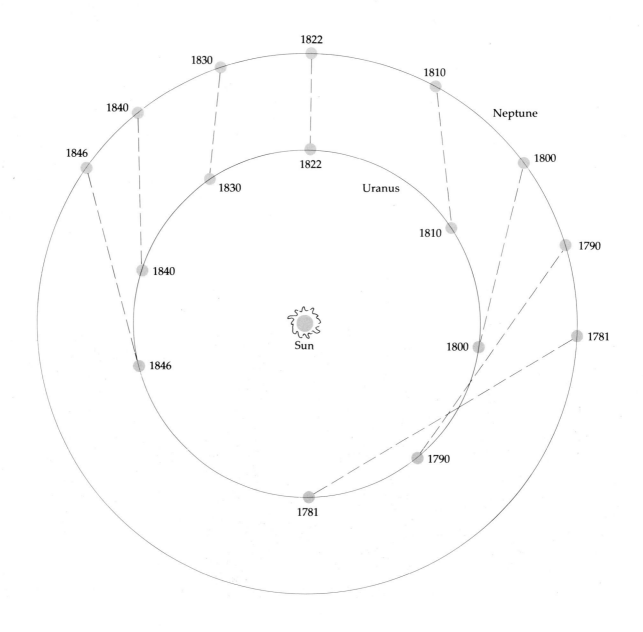

Again Airy was sceptical: 'I cannot conjecture what is the cause of these errors, but I am inclined, in the first instance, to ascribe them to some error in the perturbations. There is no error in the pure elliptic theory (as I found by examination some time ago). If it be the effect of any unseen body, it will be nearly impossible ever to find out its place.'

In this, of course, Airy was wrong, and he cannot be exonerated from blame in the famous episode which followed. In 1840 F. W. Bessel, who had been the first to measure the distance of a star (61 Cygni, in 1838) decided to attack the problem in an attempt to locate the unknown planet, but his illness, and that of his assistant Flemming, meant that he was never able to do so. The first estimated position of the new planet was worked out by John Couch Adams, of Cambridge, in 1845; he had made up his mind to tackle the problem even before his final examinations four years earlier. He wrote to James Challis, Professor of Astronomy at Cambridge, who was initially encouraging, and to Airy, who was not. No search was put in hand. Meanwhile U. J. J. Le Verrier, in France, had started work along the same lines, and in 1845 he reached a solution very nearly the same as Adams. Le Verrier's memoir reached Airy at the end of 1845. There was no telescope at Greenwich suited to the search, and so Airy contacted Challis, asking him to use the 11.75-inch refractor at Cambridge. It was 6 July 1846 before Challis began work, and even then he did so with a strange lack of energy; he had no really good star-maps of the area, and his laborious method was to chart the various fields with the object of comparing them later and seeing whether any 'star' had moved.

Le Verrier had had no success in persuading French observers to begin a hunt, and he sent his results to Johann Encke, Director of the Berlin Observatory. Two of the astronomers there, Johann Galle and Heinrich D'Arrest, started a search on 23 September 1846, and on the very first night they identified the long-expected planet. Subsequently Challis found that he had seen it twice, the second occasion being on 12 August, so that the discovery had been within his grasp if only he had taken the trouble to compare his observations. Various names for the new world were suggested, including 'Janus' and 'Oceanus', but before long Le Verrier's original proposal of 'Neptune' was accepted.

It is true that Adams was the first to give a position for Neptune; his work was almost as accurate as that of Le Verrier, who had been in error by only 55 minutes of arc. But Le Verrier and the other Continental astronomers had had no idea of Adams' work, and when the announcement was made, by Sir John Herschel, the French were deeply offended; the resulting arguments almost amounted to an international incident. Happily neither Adams nor Le Verrier took much part, and today they are recognized as the co-discoverers of Neptune, though the first to identify it were unquestionably Galle and D'Arrest.

Bode's Law breaks down completely for Neptune, which is one reason why most astronomers tend to reject the so-called Law altogether; Neptune is, after all, the third most massive planet in the Solar System, though its diameter is slightly less than that of Uranus.

Johann Galle (1812–1910), who first identified Neptune.

Heinrich Ludwig D'Arrest (1822–75), who was an assistant at Berlin Observatory in 1846 and joined Galle in the identification of Neptune.

Sir George Airy (1801–1882), Astronomer Royal at the time of the discovery of Neptune.

Left *The telescope used by Galle and D'Arrest to identify Neptune. The telescope is now in the Munich Museum.*

Below left *Predicted position of Neptune (X) and actual position (N). This was Le Verrier's position; the error was less than a degree. Adams' result was almost as accurate. This is a reproduction of a small part of the chart actually used by Galle and D'Arrest.*

Below right *The 30cm Northumberland refractor at the Institute of Astronomy (IOA), Cambridge, England, which Challis used in the search for Neptune.*

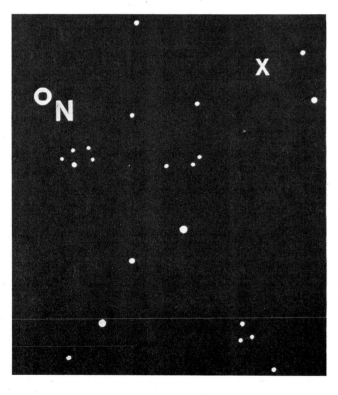

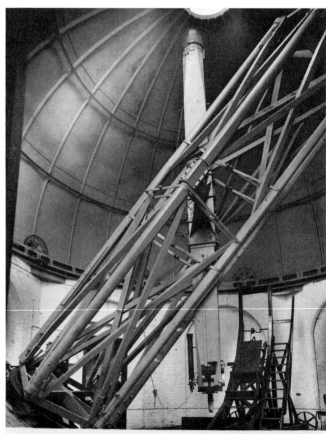

Could the Solar System be complete? Astronomers of the late nineteenth century began to have doubts. There was still something unexplained about the motions not only of Uranus, but of Neptune too. It was also pointed out that Adams and Le Verrier had been lucky inasmuch as Uranus and Neptune had been lined up in 1822 – that is to say, Neptune was then at opposition with respect to Uranus – and if Neptune had been on the far side of its orbit, the effects upon Uranus would have been inappreciable. In 1877 David Peck Todd, at the United States Naval Observatory, made a new study of the movements of Uranus, and predicted a planet at a distance of 52 astronomical units from the Sun (that is to say, about 7800 million kilometres) and with a diameter of 80 000 kilometres. He even conducted a search, using the 66-centimetre reflector with powers up to 600, hoping to find an object showing a definite disk, but the results were negative. Then, in 1879, Camille Flammarion proposed that there might be a planet moving at a distance from the Sun equal to the aphelion distances of some

Above *Percival Lowell (1855–1916).*

Left *The 24-inch Lowell refractor, used by Lowell and his assistants for planetary observations – including the hunt for the new planet.*

periodical comets, such as Halley's. T. Grigull, of Munster, predicted a Uranus-sized planet with a period of 360 years, and went so far as to give it a name: Hades.

However, the first really systematic search was instigated by Percival Lowell, who had founded a major observatory at Flagstaff and had installed a fine 61-centimetre refractor there. For his calculations Lowell preferred to concentrate upon Uranus rather than Neptune. This was because the orbit of Uranus was much the better known; the planet had completed one full revolution round the Sun since Herschel had identified it, whereas Neptune, with a period of 164.8 years, had not had time to make even half a revolution since its discovery in 1846.

Between 1905 and 1907 searches were carried out, but Lowell's Planet X, believed to have a rather eccentric orbit, a period of 288 years and a mass seven times that of the Earth, obstinately failed to show up. Neither was a second search from Flagstaff in 1914 any more fruitful, and when Lowell died in 1916 the search was given up. A quite independent prediction was made by W. H. Pickering, whose method was different from Lowell's but whose results were much the

Above *Dome of the 24-inch Lowell refractor at Flagstaff, Arizona.*

Right *Clyde Tombaugh, discoverer of Pluto. This photograph was taken in 1980 by Patrick Moore, during the symposium to celebrate the 50th anniversary of the discovery.*

same; in 1919 Milton Humason, at the Mount Wilson Observatory, made a brief photographic search in the position given by Pickering, but Planet X remained elusive.

Eventually, in 1929, astronomers at the Lowell Observatory returned to the problem. They engaged a young amateur, Clyde Tombaugh, to conduct the search, using a 33-centimetre refractor obtained specially for the purpose. Tombaugh's method was to take two plates of the same area, with an interval of several nights, and then compare them by using an ingenious measuring device known as a blink-microscope, which would show up any object which had moved during the interval. Lowell's expected planet was recorded on plates taken on 23 and 29 January 1930, and the announcement was made on 13 March, 149 years after the discovery of Uranus and, incidentally, 78 years after Lowell's birth. It was named Pluto.

Earlier searches had failed because Pluto was much smaller and fainter than had been expected. More-over, Humason had had bad luck; on examination, the image of Pluto was found on two plates, but once it was masked by an inconvenient star and on the other occasion by a flaw in the photographic emulsion.

Again all seemed well, but soon more doubts began to creep in, because Pluto's small size and unusually eccentric and inclined orbit indicated that it was unworthy of true planetary status. The discovery of Charon, in 1977, made this idea seem even more likely, and in any case there is no possibility that Pluto could perturb either Uranus or Neptune by any measurable amount; it is not massive enough. In other words, Pluto is not Lowell's Planet X. Either Tombaugh's discovery of it, not far from the calculated position, was pure chance, or else the real Planet X awaits discovery. (Let it be added that Tombaugh did not keep strictly to Lowell's prediction; he was 'working his way' round the Zodiac, so that he would certainly have found Pluto in any case.)

Various efforts have been made to see whether yet another planet can be traced. Calculations by K. Brady, based on the movements of Halley's Comet, were shown to be unsound, but it does seem definite that there is an unexplained irregularity in the movements of Uranus and Neptune, and few astronomers doubt that Planet X exists. The trouble is that is is bound to be so faint, and so slow-moving, that identification will be extremely difficult.

In 1972 one of the present authors (P.M.) suggested that something might be gained from the movements of space-craft such as Pioneers 10 and 11 and the Voyagers, which will escape from the Solar System. If contact with them is maintained well out beyond the orbits of the known planets, and any perturbations are found, Planet X might be held responsible. This, of course, is a very 'long shot'; it would depend upon Planet X being in exactly the right region at exactly the right time, but there is at least a slim chance. Otherwise, we depend upon modern charting and examination of very faint stars. Yet thanks to Voyager 2, the orbit of Uranus itself is now known better than ever before, and in time there is a possibility that Herschel's planet will show us the way to Planet X.

Even if Voyager 2 did no more, it would still rank as the most remarkable space-craft so far launched. It has given us our best views of the outer giants and the only views of Uranus; it should function for the Neptune encounter, and it may even remain in contact until it passes through the heliopause, where the Sun's effects cease to be dominant. At last we will lose track of it, probably about 2010, and for many millions of years to come it will travel unseen and unheard between the stars; in the admittedly unlikely event of its being found by an alien civilization, it might provide a clue as to its planet of origin, since it carries plaques and recordings. At least the people of Earth will never forget it. Voyager 2 has a permanent and unique place in scientific history.

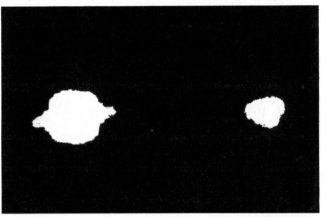

Top *The two arrows indicate the planet Pluto lying at the far (known) edge of the Solar System. The bright star is Geminorum.*

Middle *Pluto and Charon, shown separately by speckle interferometry, Pluto is to the left. The projections are instrumental effects. This was one of the first pictures to show the two bodies completely separate; it was taken at the Mauna Kea Observatory in Hawaii.*

Bottom *Halley's Comet. Suggestions that its motion is being affected by 'Planet X' have been made, but seem to be unfounded.*

Index